DESIGN

HISTORY • MAIN TRENDS • MAJOR FIGURES

by Anne Bony

For the English-language edition:

Translator
Judith Hayward

Editor
Stuart Fortey

Art consultant
Dr Richard Williams, University of Edinburgh

Series editor
Camilla Rockwood

Publishing manager
Patrick White

Prepress
Vienna Leigh

Prepress manager
Sharon McTeir

Proofreader
Harry Campbell

Originally published by Larousse as *Le Design* by Anne Bony

ISBN 0550 10194 2

Typeset by Chambers Harrap Publishers Ltd, Edinburgh
Printed in France by MAME

Introduction

How should design be defined? In what areas can it be used? Product design, industrial design, environmental design ...

The word 'design' is derived from the Latin word *designare*, which can mean to mark, trace, represent, draw, indicate, show, designate, signify, place, arrange, settle, or produce something unusual. As commonly understood, design is a discipline that sets out to harmonize the human environment, ranging from the design of everyday objects to town planning. According to the designer Roger Tallon, 'it is neither an art nor a means of expression but a methodical creative approach which can be generalized to cover all problems involving the working out of concepts'. While 'design' is an English word, it is close to the French terms 'dessiner' (to draw) and 'désigner' (to designate); its true meaning is associated with drawing and conceiving, with form and final outcome.

Design raises the whole question of how to create something when all one has are materials and procedures for transforming these materials. Designers look to give added meaning or beauty to functional objects rooted in real social contexts. From initial conception to production, they are involved in a dialectic between idea and material, and specifications take on a formal or aesthetic dimension as a result of their input. By combining usefulness and aesthetics, designers add value to objects. The debate over the distinction between what is useful and what is beautiful has been kept going since the mid-19th century by European avant-garde movements and artists.

The process of creation and innovation also takes consumers into account. Objects say a lot about human beings. They are not simply accessories of civilization; they lie at its roots. If we consider our period from an archaeological perspective, then objects constitute the most reliable materials for anthropological study. They provide information about our civilization through what they reveal about our mastery of materials, our manufacturing methods and our marketing strategies. From the primitive vase to the wheel, the object has always been a tool, a human prosthesis, and a measure of how man has come to dominate the world. It is symbolic of the great periods of history through which mankind has passed.

More than that, objects evoke a presence: they are like ghosts of the body, or indestructible witnesses, lying there like ancient monuments or relics of human creativity. In an era of mass marketing and mass consumption, the object is regarded as a set of messages which have corresponding conventional codes. In *Le Système des Objets* (1968), Jean Baudrillard describes this relationship between social setting and the symbolic combinations of objects. His analysis is based on semiology – the science of signs; he classifies objects according to a system which is comparable to that of language.

Design arose from the desire to create a synthesis between form and function and apply it to industrial objects. Its emergence coincided exactly with the Industrial Revolution. A combination of social, economic and production factors then led to unprecedentedly rapid development. Design realized its full potential when utilitarian production, enhanced with the aesthetic values that design brought to it, was able to meet the demands of industrial production. It then found its way into the area of domestic appliances, where its aim was to reconcile attractiveness and usefulness, both in planning and practice. Design became a real player in the social stakes, an active element in the production-consumption system.

Design has evolved over the course of history. Its role has been crucial from the time of the first World Fair (the Great Exhibition held in London in 1851), when products of the machine age were shown alongside masterpieces of traditional craftsmanship, until today, when interactive design allows a dialogue between the individual and the computer and the creation of a mind-boggling virtual reality – which raises the question: is an object still an object in a virtual world?

This book charts the ever-changing historical development of design over the last 150 years, during which time theories have been constantly re-examined and renewed: from the 19th-century Arts and Crafts movement in England and the Bauhaus in Germany via the birth of Industrial Design in the USA, industrial aesthetics in France and the Italian miracle, to the 'everything in plastic' of the Pop years; from the anti-design of the Alchymia and Memphis studios to virtual design at the start of the 21st century.

The history of design touches on many areas: aesthetics, sociology and politics; technology and materials; and commerce and the production-consumption system.

Anne Bony

Contents

Contemporary Design (1939-58) 84

The Pop years (1958-72) 120

Alternative design (1973-81) 148

The age of aesthetics (1851-1914)

- The Great Exhibition of 1851
- The English Arts and Crafts movement (1862)
- Educating the public
- Creative engineering
- Art Nouveau: a new attitude
- French Art Nouveau
- Stylized forms: the Glasgow School
- The Vienna Secession: 'Only that which is functional can ever be beautiful'
- A community experiment: the Darmstadt Artists' Colony (1899–1914)
- The German Secession

This was a time of struggle and paradox, generating a desire (albeit illusory) for a symbiosis between industry and art; a time when a great many contradictory trends coexisted. Based on scientific progress and its practical applications, the Industrial Revolution had seen the emergence of new techniques and new materials. Saint-Simon, Owen, Fourier, Proudhon and Marx advocated a socialist society and egalitarian models. The American architect Louis Sullivan developed a functionalist ideology: 'Form follows function'. John Ruskin and William Morris championed craftsmanship and a nostalgia for beautiful, well-crafted work. Nature was looked at afresh, and a wide variety of organic forms made their appearance in the design of interiors and objects – a stylized approach which had an impact on the assumptions underlying abstraction.

The Great Exhibition of 1851

Arising from the desire of the industrial powers to display their diverse outputs in a peaceful context, World Fairs are valuable points of reference for studying the decorative arts of the second half of the 19th century. They presented the work of factory owners and manufacturers keen to display their finest products.

The 1851 Great Exhibition in London was the first major industrial and technological event of the 19th century and an unparalleled endorsement of human industry. Henry Cole, who promoted the exhibition, went so far as to claim that it was, in a way, a 'history of the world'. Until then, local or national exhibitions had been held regularly. In 1849, Henry Cole had visited one of these – the quinquennial exhibition in Paris – with the critic Matthew Digby Wyatt, and on his return to London in June he proposed to Prince Albert at Buckingham Palace the idea of staging the first international exhibition in the British capital. The Prince Consort agreed, becoming personally involved in helping him make the scheme a reality. Henry Cole was endowed with a great deal of practical ability, and the organization of the event turned out to be a great success, both financially and in terms of prestige. It made a profit of £186,000, which was allocated to industrial research.

Previous page: Interior of the drawing-room at the Hill House, Charles Rennie Mackintosh, 1903. The whole interior is suffused with light: white painted wooden furniture and armchairs in natural wood, including a high-backed one with Celtic motifs. The house was built at Helensburgh near Glasgow for the publisher William Blackie. Hunterian Museum, Glasgow.

Crystal Palace

The design eventually chosen for the exhibition building was much debated and aroused fierce controversy. It was thought up in nine days by Joseph Paxton (1801–65), adviser to the Duke of Devonshire. Neither architect, engineer or scientist, he proposed a structure that ressembled a large greenhouse, a giant version of the one at Chatsworth. It was christened the 'Crystal Palace' some months later by the magazine *Punch*. Its construction was based on sound principles, involving pre-

Joseph Paxton's Crystal Palace, the building housing the 1851 Great Exhibition in London. Made from prefabricated elements, it was 563m long, 124m wide and 36m high; it was made up of 2,224 girders and 300,000 glass elements. Constructed in six months – an outstanding technical achievement – it served as a model for the Gare de l'Est in Paris.

fabricated elements which were mass-produced and then assembled. The first column was erected on 26 September 1850 in Hyde Park. The main structure was completed within four months. One of the stipulations made by the exhibition commission was that trees should be protected: three existing large elms were included within the structure, which gives some idea of the height of the vaulted transept. News of the building of the palace and the exhibition became more widely known, and from February onwards Londoners saw manufactured goods produced by every known process arriving from the four corners of the world. For the first time ever, the arts, sciences and industry were being brought together. This development of the applied arts was an endorsement of positivism, which saw science as an opportunity to bring material well-being to all social classes. The west side of Crystal Palace was allocated to the display of products from Britain and its Empire, while foreign contributors occupied the east side. The organizers worked out various national and general categories that formed the basis for the juries' reports in the competitions. The main divisions were as follows: A, raw materials; B, machinery; C, manufactured products: textiles; D, manufactured products: metal, glass and ceramics; E, miscellaneous; F, fine arts. The criteria taken into account by the juries in awarding the medals were: novelty; economy in manufacture and maintenance; strength; quality of execution; suitability for purpose; innovativeness in re-examining established principles; progressiveness as regards formal aesthetics; precision and reliability of performance; beauty of drawing, form or colour; and form and colour with reference to utility. The highest award was the Council Medal. Medals were stamped with the effigies of Queen Victoria and Prince Albert.

The marriage of art and industry

The exhibition was officially opened by Queen Victoria on 1 May before a crowd of 25,000 people. *The Times* (2 May 1851) described the ceremony: 'There were many people there who were used to magnificent spectacles, many who had been present at coronations, festivals and solemn occasions; but they had never seen anything to compare with this ... Round them, among them and above them, what was beautiful and useful in nature and art was presented ... Some saw it as a second, more glorious celebration of royalty'.

The exhibition put on display not only utilitarian objects but also artistic masterpieces. It is ironic that the works selected, although modern in spirit insofar as the artists and craftsmen made use of manufacturing processes, still harked back to the past. Each country chose the style that reflected its national identity to its best advantage: for example, Italy presented items inspired by the 15th-century Renaissance, and France opted for the 16th-century Renaissance and the period of Louis XV.

England brought the Gothic style back to life with a section called the 'Medieval Court', comprising a large number of decorative objects. This was designed by Augustus Welby Northmore Pugin, the hero of the Gothic revival. He examined closely the rules of Gothic architecture and concluded that the structure inherent in each created work should be made visible. He offered clarity as opposed to the deceptions of eclecticism.

France exhibited a masterpiece in the form of the Duchess of Parma's dressing table and toilet set by François-Désiré Froment-Meurice, who was the head of a factory employing over a hundred workers at that time. This piece of furniture was a resounding success, receiving the Council Medal. It was the result of a truly collaborative venture, involving the goldsmith Froment-Meurice, the sculptors Jean-Jacques Feuchère and Adolphe-Victor Geoffroy-Decheaume, the draughtsman Marie-Joseph Liénard, and the enamellers Sollier, Grisée and Meyer-Heine. It was a remarkable work involving the harmonious co-operation of various artists, but can it really be regarded as a synthesis of the arts?

Close collaboration between manufacturer and artist was necessary. This was emphasized by Count Léon de Laborde (1807–69), author of a massive report on the Great Exhibition with the title *De l'Union de l'Art et de l'Industrie* (1856). By securing the collaboration of artists, some industrial factories did clearly improve the quality of their output. Manufacturers had become capable of producing items that demonstrated great technical virtuosity and that were in no way inferior to the

Michael Thonet

In 1860, the firm of Thonet produced *Rocking Chair no.1*, made of bent beechwood. The company adapted the structural elements of a wrought-iron armchair, made by R W Winfield & Co. of Birmingham, which they had seen at the 1851 Great Exhibition in London. The bentwood technique is used to optimum effect in this archetypal rocking chair.

Thonet's output is an example of innovation keeping pace with industrial progress. In 1830–5, the German Michael Thonet (1796–1871) perfected a process for bending wood. He obtained an imperial charter in 1842, opened a shop in Vienna in 1852, established his company in 1853 and lodged an exclusive patent in 1856. The items he created are perfect examples of practical furniture design using forms that could only be produced industrially: for example, his *Chair no. 14*, whose design is reminiscent of seats from the Biedermeier period. The chair consists of just six pieces: the backrest and back legs in one piece; the interior of the backrest, which is screwed on; the seat; the two front legs; and a ring to stabilize the four feet.

This chair fulfilled the optimum conditions for mass production and export. A box measuring approximately one cubic metre could hold 36 unassembled chairs, which were then put together at the delivery site. In order to make mass production feasible and cut costs, Thonet had the idea of making the pieces interchangeable between a large variety of models. In London in 1851, he exhibited a range of seats based on the same model: an upright chair, an armchair and a bench. In 1859, a poster served as his first illustrated sales catalogue, depicting 26 models in various shades and colours. The items were numbered, thus making matters easier for the buyer.

Between 1841 and 1900, Michael Thonet took part in 55 exhibitions, 13 of which were World Fairs. He produced sales catalogues in several languages. Being totally without ornament, Thonet's furniture was acceptable in public places, but could not gain entry into middle-class homes. So, in the 1880s, the manufacturer found himself forced to introduce the so-called Makart style into his production line to satisfy middle-class taste.

work produced in earlier centuries. And research in all fields was resulting in the invention of machinery that made it possible to reproduce plentiful, cheap models that had formerly been made by hand. Woodcarving, for example, could now be done by machine tools. In gold- and silverware, the perfecting of electrochemical processes led to the disappearance of goods plated in the traditional manner. As for furniture, the invention of plywood and the use of fast, precise new machinery (circular saws and spindle-moulding machines) made it possible to imitate 17th-century cabinet work.

But not everyone understood what was required. Some industrialists were uncomfortable with the new materials and new techniques, and merely copied the styles of the past without taking into account the specific nature of the techniques involved. Thus processes aiming to imitate the most sophisticated materials such as tortoiseshell or cordovan proliferated, resulting in an inevitable decline, mediocre copies and shoddy merchandise. The effect was disastrous: pastiche and eclecticism flourished.

The champions of traditional craftsmen put forward a different point of view, seeing craftsmen as the only people capable of producing not only beautiful goods but worthwhile ones. They castigated industry, which was incapable of producing uniquely beautiful objects and which subjected workers to virtual slavery.

In 1851, Matthew Digby Wyatt published *Metalwork*, in which he attacked 'utilitarianists' and 'idealists'. The former were guilty in his eyes of producing ugliness in our daily lives; the latter of 'sacrificing comfort and convenience to ornament and effect'. The 1851 Great Exhibition stimulated a thoroughgoing, hard-fought debate about the role of ornament and the place of industrial processes in design. The exhibition also encouraged trade and launched a new trend: tourism and the discovery of new countries.

The English Arts and Crafts movement (1862)

The period between 1850 and 1875 in England was characterized by a number of new aesthetic elements: the emergence of the Pre-Raphaelite group, William Morris's Arts and Crafts movement, a new style in architecture, the Japanese influence, the Aesthetic Movement, and seeing nature as a source of inspiration. These various trends gave rise to a renaissance of the decorative arts, to a heightened social awareness, and to the ambition that artists and craftsmen might work together to educate public taste and so improve the living environment.

The Arts and Crafts Exhibition Society was established in 1888 (at the height of the movement's activities) in order to spread its message. The idea of a regular exhibition to promote furniture design and the decorative arts had been envisaged by Walter Crane as early as 1886.

Charles Robert Ashbee (1863–1942), himself an architect, set up the Guild of Handicraft in 1888 along with craftsmen such as John Pearson and John Williams. They produced objects in a variety of materials – leather, wood, base and precious metals – and carried out decorative work. In 1898, the guild was commissioned to make the furniture for the palace of the Grand Duke of Hesse at Darmstadt, designed by Mackay Hugh Baillie Scott.

Ashbee was a central figure in the Arts and Crafts movement, passionately interested in Ruskin's ideas. He went to the USA, where he met Frank Lloyd Wright. In 1896, he was elected a member of the Art Workers Guild, and in 1898 he founded the Essex House Press. The Guild of Handicraft opened a shop in London's Bond Street in 1899. The organization spread internationally until it was dissolved in 1907. In his book *Craftsmanship in Competitive Industry* (1909), Ashbee set out the principles underlying his work, happily accepting the use of machinery in the production process.

William Arthur Smith Benson (1854–1924) was a metalworker and a member of the Art Workers Guild. He took over as managing director of William Morris's firm when Morris died in 1896. Remaining loyal to the

Secretaire designed by Charles Robert Ashbee, 1902. Made by the Guild of Handicraft, it is in mahogany, with ebony and holly veneer; the base is made of oak, painted red, and the inlays are wrought iron.

William Morris

Interior decoration of the Red House, 1859. In the green dining-room the panels, ceiling and frieze were designed by Philip Webb. William Morris, assisted by his fellow-workers at William Morris & Co., promoted the idea of total design. London, Victoria and Albert Museum.

William Morris (1834–96) was the theorist of the Arts and Crafts movement. A disciple of John Ruskin (1819–1900), he adhered to ideas that Ruskin had expressed in *The Stones of Venice* (1851). In the chapter 'The Nature of Gothic', Ruskin describes the qualities of the Gothic style, including its 'honesty' on the aesthetic and moral level, and demands a return to the dignity of craftsmanship. William Morris met Philip Webb (1831–1915), a young architect, and in 1859 commissioned him to build his house – the Red House at Bexleyheath in Kent – while he himself assumed responsibility for the interior decoration. This experience encouraged him to create the firm of Morris, Marshall, Faulkner & Co. in 1861, a move made possible by his private fortune and the support of his friends, the Pre-Raphaelite painters Edward Burne-Jones (1833–98) and Dante Gabriel Rossetti (1828–82). In 1875, he took over sole direction of the business, which was then renamed Morris & Co. The workshops at Merton Abbey were enlarged to allow the production of wallpapers and printed fabrics and the weaving of carpets and tapestries.

Work was organized on a communal basis, and the workshop also served as a sales outlet. The advertising leaflets stated: 'A society of artists has just been formed with the aim of producing objects distinguished by an artistic character and selling at high prices; they have resolved to devote themselves to producing useful objects and it is their intention to lend them artistic value.' In 1891, William Morris set up a printworks, the Kelmscott Press, not far from his London home. He himself designed the typefaces, initial letters and borders and decided on the layout of the pages, while the illustration of the nearly 70 works published between 1891 and 1896 was entrusted to Edward Burne-Jones.

William Morris was one of the heirs of Gothic revival, especially in the field of furniture. He favoured simple, unconcealed structures, even if he or his Pre-Raphaelite friends did paint some of the furniture. He advocated the use of natural materials, and his love of fine handicraft practised within workshops was reminiscent of the medieval guilds. His desire to bring artists and craftsmen together as closely as possible and his search for a unified concept of architecture and interior decoration derived from his fascination with a mythical Gothic society as imagined by the visionary John Ruskin. However, Morris's work differs from the Neo-Gothic in its stylization: an organic vitality and tangled density of motifs that were probably inspired by rich Italian textiles of the 15th and 16th centuries.

Orchard wallpaper, designed by John Henry Dearle for William Morris & Co., 1899.

Kelmscott cabinet, designed by Charles Francis Annesley Voysey, c.1890, oak with wrought-iron mounts with nature-inspired motifs; made by F Coote. London, The Fine Art Society Collection.

ideas of Morris and Ruskin, he produced furniture, wallpapers and wrought-iron railings. His metalwork designs were sold by Samuel Bing in Paris. But gradually his views altered, and he allowed the forms of light fittings, for example, to be dictated by the requirements of industry. He confirmed this commitment to industry by equipping his new factory in Hammersmith with the best machinery available.

Another interesting figure is Christopher Dresser (1834–1904). Although central to the Victorian period, he looked towards the future. A pupil of the Government School of Design, Dresser favoured highly geometrized naturalistic ornamentation and was responsive to technical progress. He understood the nature of industrial manufacturing processes, accepted mechanization and was aware of the need for products specifically adapted to it. He forged links with businesses to produce his designs, thus demonstrating that work done by machines was not necessarily at odds with creative design. From the late 1870s onwards, he designed everyday objects that could be manufactured in brass, or more simply still in brightly coloured enamelled iron, on behalf of large factories such as Elkington, Hukin & Heath and Dixon & Son. Like Edward William Godwin (1833–86), he was impressed by the Japanese stand at the 1862 World Fair in London and made an extended visit to Japan in 1876–7, which inspired him to use more restrained forms in his work.

Secretaire designed by Arthur Heygate Mackmurdo, c.1886. This piece, severe and geometric in form, is a forerunner of Mackintosh's style. London, William Morris Gallery collection.

In the 1880s, the Arts and Crafts movement attracted designers with an idealistic approach. Interested above all in methods of working, the movement developed an attitude towards life rather than a style. It was emulated in the USA, Scandinavia and Central Europe in particular. Its impact on the aesthetic front was considerable, especially on Art Nouveau.

Educating the public

In the 19th and 20th centuries, World Fairs assumed the role that royal courts had played in the 18th century, giving their names to the style of the period. Exhibitions, whether international, national or thematic, were a 19th-century phenomenon, motivated by industrial growth in a society with a rising population – a society that had to be informed and stimulated.

In France, the first exhibition devoted to 'industrial products' was held on the Champ-de-Mars in August 1798. Featuring only French products, it set out to prove that independent workers could make objects as fine as those produced by trade guilds. Successive governments encouraged industry to develop a close relationship with science and art, seeing this as one of the prerequisites of progress.

National exhibitions in Paris came thick and fast: in 1801, 1802, 1806, 1819, 1823, 1834, 1839, 1844 and 1849. Throughout this period the predominant style was pastiche, indicating a lack of creativity. Following on from the 1851 Great Exhibition in London, Paris organized a World Fair in 1855. Attention was focused on the Salle des Machines (Machinery Hall) and the Palais de l'Industrie (Palace of Industry). In

1867, the World Fair was again held in Paris, where exhibits from Japan impressed creative artists such as Eugène Rousseau, the glassmaker Émile Gallé, many interior designers and the goldsmith Christofle. In 1876, a World Fair was organized in Philadelphia to celebrate the centenary of America's independence. The 1878 exhibition in France confirmed the reputation of Art Nouveau, with a display of jewellery by Tiffany & Co. from New York and an important interior by Émile Gallé. The 1889 Paris World Fair marked a turning point: artists and designers finally broke free from a dependence on past styles thanks to techniques which allowed them to think the impossible. The Eiffel Tower is a monument to this boldness of vision. Visits to exhibitions in other countries encouraged development and invention, as Louis Sullivan and the Chicago School demonstrated at the World's Columbian Exhibition in 1893. Plans for the 1900 exhibition in Paris were drawn up very early. 'All sectors of human activity will derive equal benefit from this huge exhibition, and a bright light will be shed on the moral and material conditions of contemporary society,' wrote Jules Roche, the French Minister for Trade and Industry at the time. The 1900 World Fair, the culmination of the previous thirty years, marks both the peak and the decline of Art Nouveau.

Museums and magazines

The South Kensington Museum in London (now the Victoria and Albert Museum) came into existence immediately after the Great Exhibition of 1851, bringing together an exemplary collection of items. In Vienna, a Museum of Arts and Industry opened its doors in 1864. Most towns in Germany and Austria had associations for the decorative arts ('Kunstgewerbevereine'); these bought works, set up schools and published specialist magazines. From the time it opened in France in 1877, the Musée des Arts Décoratifs had a policy of exhibiting and acquiring both contemporary and older works. In 1884, the Union Centrale des Beaux-arts Appliqués à l'Industrie merged with the Société du Musée des Arts Décoratifs to become the Union Centrale des Arts Décoratifs. This mounted exhibitions with special themes, such as clothing and tapestry. Then, starting in 1880, it staged the first exhibitions of technology: metalworking, wood, paper and textile arts; in 1884, stone, wood, earthenware, glass and ceramics; and in 1887, the Union put on a wide-ranging exhibition called 'Les Arts Appliqués à l'Industrie'. Between 1889 and 1900, salons that had traditionally been the preserve of painting – such as the Salon de la Société Nationale des Beaux-arts and the Salon des Artistes Français – gradually opened their doors to the decorative arts.

The exhibitions and salons brought the public into contact with new objects, while art journals kept them informed, providing illustrations and criticism of innovative forms. The first publication was *La Revue des Arts Décoratifs*, brought out by the Union Centrale des Arts Décoratifs

The Eiffel Tower, built by the engineer Gustave Eiffel for the 1889 World Fair in Paris, is a remarkable technical achievement. It has a highly visible structural logic.

in Paris. *The Studio* was launched in London in 1893, followed by L'Art Moderne in Brussels, which promoted the Brussels avant-garde. *Dekorative Kunst* ('Decorative Art') was first published in Munich in October 1897, while a French edition, *L'Art Décoratif*, was published in Paris the following year, helping to form an artistic link between France and Germany. The monthly *Art et Décoration* first appeared in 1897. These publications provide the best picture of the decorative arts of the time.

Creative engineering

There is an ambiguity inherent in the idea of creative engineering: it is seen as desirable and yet can also be seen as undesirable; it is a source of both inspiration and repulsion. Engineers have always worked closely with architects, and the 19th century saw the beginning of a reconciliation between the notions of art and technology. Although engineers do not set out to create beauty as such, true creativity can indeed be expressed in construction and civil engineering projects.

At the end of the 19th century, there were many examples of buildings with metal structural frames in the USA. William Le Baron Jenney erected the first skyscraper for the Home Insurance Company in Chicago in 1883. It was the first construction entirely in metal which had been based on research into structural principles. The tower erected by Gustave Eiffel and the Salle des Machines built by the architect Ferdinand Dutert and the engineer Victor Contamin for the 1889 World Fair represented the high-water mark of creative engineering. One of the most striking aspects is the increasingly wide range of materials that were used.

The cast iron and the iron of the first Industrial Revolution were followed by steel and reinforced concrete, while traditional materials such as glass and brick began to be produced industrially, thus leading to an ever-increasing number of products. Industrialization and the standardization that went with it did not necessarily mean a reduction in the range of building elements available – on the contrary. Although the dimensions of bricks, for example, were standardized towards the end of the 19th century, at the same time a whole series of derived products (including enamelled brick) appeared on the scene.

Building materials became more diversified in the 20th century due to the introduction of metal alloys and concretes, the development of float glass and the perfecting of plastics and composite and intelligent

Bed designed by Gustave Serrurier-Bovy, 1899; mahogany with brass inlays, with taut curved lines, decorated with patterned embroidery work in silk.

materials. These opened up new possibilities. At the beginning of the 19th century, however, the choice of materials was still limited.

Art Nouveau: a new attitude

Art Nouveau owes much to William Morris. He fiercely challenged the aesthetic values of the mid-Victorian era, condemning the use of machinery and the wrong-headed division of labour with its resultant loss of humanity. Also underpinning the movement was Viollet-le-Duc's thinking about the internal logic of Gothic art, as set out in his *Dictionnaire raisonné de l'architecture française du XIe au XVIe siècle* (1854– 68). This revelation of its internal structure, its codes and its general application would influence Art Nouveau designers. Yet another source of inspiration was the colonial conquest of the East, which opened the eyes of London, Paris and Brussels to an unfamiliar artistic world. In London, Liberty & Co. opened a shop selling oriental imported goods. In Paris, Samuel Bing specialized in oriental prints before setting up his own shop, itself called 'Art Nouveau'. Gustave Serrurier-Bovy opened a decorative goods shop in 1884 in Liège, importing pieces from Japan and the East. This contact with the Far East, which influenced Van Gogh and Gauguin, led artists to look at nature in a different way. Social behaviour was also undergoing profound changes, including a democratization of the arts.

An attitude rather than a style, Art Nouveau was a movement with one main aim: to overturn the established order in the fields of fine art and the applied arts. No single architect, designer or school is representative of Art Nouveau; each sought to come to terms in their own way with the prevailing dependence on the art of the past. Stylistically speaking, exuberance was the main feature in France, Belgium and Germany, while in Scotland and Austria restraint, even austerity, were the dominant tendencies. The solution suggested by Art Nouveau for modern decorative interiors involved harmonizing all the elements in a room, from the general colours down to the most minute detail of the tiniest object, such as a keyhole cover or a hinge on a piece of furniture. It was, in short, a concept of total design.

Art Nouveau in Belgium

The reign of the Belgian king Leopold II provided a prosperous context for the development of Art Nouveau. Belgium was a young country, at the forefront of industry, research and trade, and with a well-to-do, generous middle class that championed social causes. It was for a time the dynamic melting-pot of the new style, acting as a bridge between England and continental Europe.

The Aesthetic Movement gave England a head start over the Continent in the search for a new style; however, it was in Belgium that

Dining-room in the Maison Horta, Brussels, 1898. The house was built by Horta for himself, using radical new designs. The dining-room is decorated with ceramic tiles, bas-reliefs and lights in the shape of flowers. The house is now the Horta Museum.

the new approach came to fruition. Brussels was a vitally important centre for the propagation of new ideas. The group Les Vingt or Les XX (The Twenty), founded in 1884 under the aegis of the Brussels lawyer Octave Maus, had its origins in the journal *L'Art Moderne*. Its aim was to attract progressive European artists who would challenge the prevalent dependence on outdated styles. The first of their annual exhibitions of avant-garde European works featured Auguste Rodin and created a sensation. The fact that Toorop, Ensor, Khnopff, Van de Velde and Van Rysselberghe were among the signatory members of the group's charter made it possible, in 1892, for decorative works of art to be exhibited on an equal footing with painting. Les Vingt provided a forum for all those excluded from official competitions. The stylistically revolutionary works of Mackmurdo were first exhibited in Brussels, where they were admired by the artists resident in the capital.

Exhibitions were held every year, until the group was disbanded in 1894. It reformed under a different name, La Libre Esthétique, and continued to support and encourage the decorative arts. The annual exhibition was a showcase for the flourishing state of all the arts: painting, sculpture, graphic art and the applied arts. After a long battle against the idea of a hierarchy of genres, the applied arts finally seemed to have achieved parity with the fine arts. The architects Horta, Van de Velde and Serrurier-Bovy were the first to produce furniture designed to fit in harmoniously with the buildings containing it. Paul Hankar, Georges Hobé, Antoine Pompe and Georges Lemmen would follow suit; their work was exhibited at the Salon des Vingt.

The work of architects

In 1893, the engineer Émile Tassel asked the architect **Victor Horta (1861–1947)** to build a private house for him in Brussels. In fulfilling this commission, Horta revolutionized the design of the traditional middle-class house by the conspicuous use of iron and cast iron. He linked structure and interior decoration by creating a decorative language based on the arabesque. Horta hung a glazed lantern light above the stairwell of the house, enclosing the space itself in flowing ribbons that wind around one another and rise like random flames from the bottom of the stairs. The ribbon motif is repeated in parallel on the wrought-iron banisters, in the painted arabesques on the ceiling and in the mosaic arabesques on the floor. The interior decoration has a remarkable unity, even in the smallest details, with the architectural line being carried over to the furniture. The pieces of furniture that Horta designed were customized for each client; none of his designs was intended for industrial production.

His rejection of stylistic conformism and his friendships with members of the Belgian Popular Party influenced the choices Horta made when building the Maison du Peuple in Brussels. He chose brick, stone, iron and cast iron. His reputation was greatly boosted by a collection of furniture and murals (belonging to an interior scheme for the Hôtel Solvay and the Hôtel Van Eetvelde) which he presented at the 'La Libre Esthétique' exhibition of 1897. Nevertheless, unlike Van de Velde and Serrurier-Bovy, he had no intention of mass-producing his furniture.

Gustave Serrurier-Bovy (1858–1910). The work of this architect and furniture designer helped to forge links between Belgium and England. Early in his career he travelled to England to attend craft courses, returning with ideas for simple, well-constructed furniture. At his shop in Liège he sold his own furniture as well as English furniture and was granted a franchise to sell goods made by Liberty. In 1894, he was the first furniture designer to show his work at the 'La Libre Esthétique' exhibition. He set up a complete interior, inspired by the English Neo-Gothic style, and in a room at the Musée des Beaux-Arts he installed a decorative scheme complete with furniture, curtains, wallpaper and lights. The following year he exhibited a 'craftsman's room' with a view to promoting 'the popularization of the aesthetic sense'. Henry van de Velde and, later, Horta also showed interiors at the salon. Serrurier-Bovy's furniture is less exuberant than Horta's, but emphasizes the tension lines in its structure. That visual tension is a feature picked up by Van de Velde, who rightly regarded Serrurier-Bovy as a seminal influence on Belgian design.

Serrurier-Bovy was one of the founders of the 'L'Œuvre Artistique' exhibition – devoted exclusively to the decorative arts – that opened in Liège in 1895. Works by the Glasgow School and the first projects by a young French architect, Hector Guimard, were shown there. Serrurier-Bovy opened his shop in Brussels, then established a branch in Paris with the name 'L'Art dans l'Habitation' (Art in the Home). Greatly impressed by his visit to the exhibition staged by the Artists' Colony in Darmstadt, he abandoned curves for more geometric lines – a simplification of forms that reflected his desire to reduce the cost of furniture and household objects. For a competition for decorating and furnishing homes inexpensively, he entered a piece of furniture called *Silex*, made of elm and poplar. This was supposed to be assembled by the purchaser and decorated with stencils if they so wished. After a few difficult years, his business recovered, as his restrained but elegant furniture at the 1910 exhibition in Brussels demonstrated. Serrurier-Bovy died suddenly in November 1910.

Rejecting the hierarchy of the arts

Painters and sculptors began to make objects for the home. The painter Willy Finch took up ceramics and Fernand Dubois took up sculpture, while others exhibited statuettes, candelabra and table centrepieces. Van de Velde, a painter, made his views on the decorative arts clear: 'We cannot countenance a division that is determined to classify the arts by

Henry Van de Velde

Interior by Henry van de Velde, designed in 1898–9. The large desk is made of oak; the frieze, lamps and handles are in gilded bronze. The matching armchair is made of oak and leather. On the walls there are decorative panels painted by Pierre Bonnard in 1891: *Woman in a Polka-Dot Dress, Woman in a Checked Dress, Woman in a Blue Pelerine,* and *Sitting Woman with a Cat.* Paris, Musée d'Orsay.

Henry van de Velde (1863–1957), a painter, furniture designer and architect, belonged to Les Vingt. Motivated by the social upheaval sweeping through Europe, he drew inspiration from English art and the activities of Ruskin and Morris, concentrating on new concepts of form and dreaming up a completely new approach to the applied arts. He set out his manifesto *L'Art Futur: Déblaiement d'Art* (The Future of Art: Clearing Out Art) at the first 'La Libre Esthétique' Salon in 1894. The following year he designed his first house, Bloemenwerf, in a suburb of Brussels. Its furniture was restrained, and its decoration harmonized with the movement of its structural lines. Much interest was generated and Van de Velde found himself at the centre of an international movement. Julius Meier-Graefe introduced him to Samuel Bing in 1895, and he designed four interiors for Bing's new shop 'L'Art Nouveau' in Paris. The shop became the focus of a movement which eventually adopted its name.

Van de Velde achieved great success at the Dresden arts and crafts exhibition of 1897. Meier-Graefe introduced him to the journal *Pan*, for which he wrote an article on the design and manufacture of modern furniture that aroused widespread interest. He set up a factory for the commercialization of 'the arts of industry, construction and ornamentation' at Ixelles-lez-Bruxelles in 1898. The commissions Van de Velde received in Belgium were poor compensation for a critical hostility towards him, and he eventually accepted an invitation to give a series of lectures in Germany in 1900. The Grand Duke of Saxe-Weimar asked him to found a school of decorative arts in 1908 – a sign of official recognition for this advocate of renewal and change. He introduced an innovative teaching system based on the direct cultivation of sensibility and the constant use of inventiveness, uninfluenced by past models. His methods were adopted by Walter Gropius in 1919, forming the basis for the Bauhaus. The buildings for the Weimar school designed by Gropius are powerful expressions of Van de Velde's ideas.

order of importance, a separation of the arts into fine arts and second-class arts, in other words minor industrial art.' Non-traditional materials were being used at the time: for example, dark woods from the Congo were married with worked metal. In Belgium, the Art Nouveau movement lasted between 15 and 20 years. It came to an end with the Turin exhibition of 1902 and the Brussels exhibition of 1905. The curves became less pronounced, and objects became more geometric; the Viennese influence can definitely be detected in the Palais Stoclet, built by Josef Hoffmann in Brussels in 1905.

French Art Nouveau

Nineteenth-century France was a backward-looking century: the styles in fashion were recycled from various periods of the past. This met with the disapproval of architects, who developed an organic style, creating expressive façades which were in perfect harmony with the interiors. Writing about modern furniture in *La Revue d'Art* in 1899, the architect and art critic Francis Jourdain described the collaboration that had finally developed between architect, artist, sculptor, engraver, musician, writer and decorator: 'All share the same vision, the same common aesthetic aim and the same ideal'. One of the greatest successes of Art Nouveau was its holistic approach to furniture. Before 1900, items of furniture were usually designed individually and then fitted into the typical 19th-century interior with a total disregard for harmony. While it might have been possible to buy a set of chairs, people had to go to extra trouble to find the stools, tables and sideboards needed to complete the furnishing of a reception room. Traditional cabinet-makers did not look on Art Nouveau with favour; it went against their basic belief that a piece of furniture had first and foremost to be well designed and functional, and only when that had been achieved could it be decorated. The new movement gave priority to decoration, however – indeed at times going too far in that direction and producing extravagant, unfunctional items.

The desire to bring together different areas of art to create a 'modern' interior was also reflected in the way galleries were arranged. The official opening of the dealer Samuel Bing's Parisian gallery 'L'Art Nouveau' was one of the most momentous events in the history of the movement. Works produced by a whole variety of techniques were displayed together in spaces resembling private interiors. In these spaces, designed for the most part by Van de Velde, Bing presented decorative ensembles that he had commissioned from painters. Paul Ranson decorated the walls of a dining-room, Maurice Denis designed an entire bedroom, and Théo van Rysselberghe created a fireplace. But Bing's most original idea was to commission from the Nabis and Henri de Toulouse-Lautrec a series of sketches for small domestic stained-glass panels, which were then made in Tiffany's workshops in New York. In Paris,

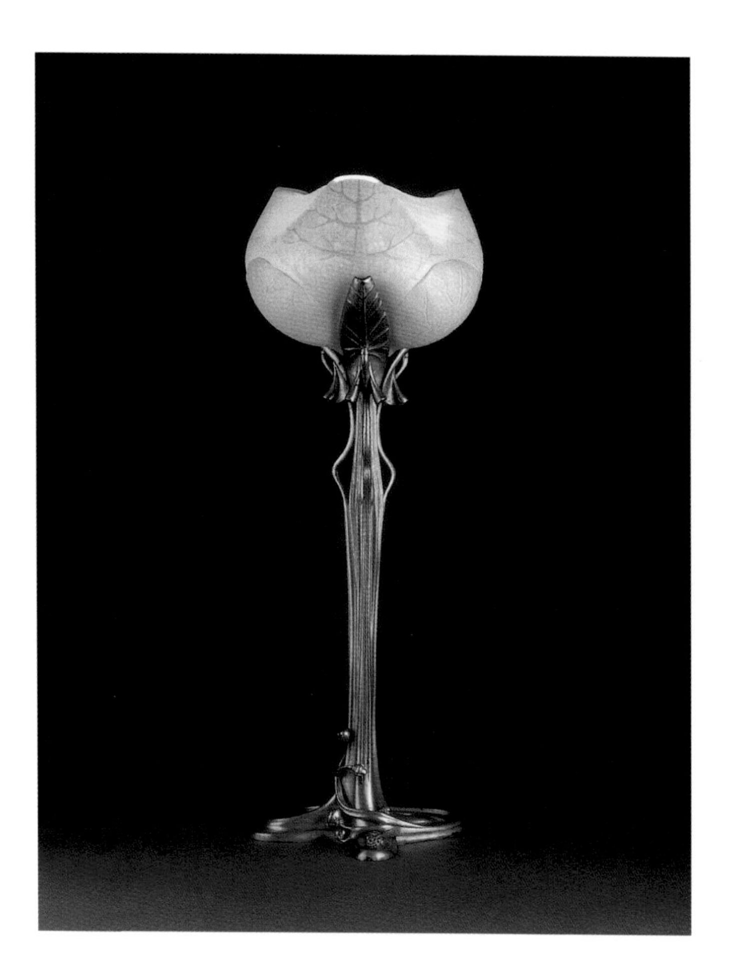

Water Lily table lamp by Louis Majorelle, bronze and blown glass, c.1902.

Gérard Soulié promoted the movement in his reviews of the annual exhibitions. The Goncourt brothers, Roger Marx and Émile Nicolas – all of them eminent art critics – were also enthusiastic supporters of the style.

The Nancy School

In France, the Nancy School was pre-eminent in the design and manufacture of Art Nouveau furniture. It brought together a group of very talented designers, including Louis Majorelle, Émile Gallé, Jacques Gruber, Eugène Vallin and later, from the younger generation, Camille Gauthier, Henri Hamm, Louis Hestaux, Laurent Neiss and Justin Ferez. The Art Nouveau architect Émile André also designed a range of furniture that harmonized with his buildings. Victor Prouvé collaborated extensively with him, offering technical advice, while craftsmen took care of the decorative accessories: Alfred Finot, Ernest Bussière and Ernest Wittmann specialized in ornamental sculptures, Charles Fridrich and Fernand Courtex in fabrics and textiles, and the Daum brothers in glass.

The main characteristic of Art Nouveau in Nancy was its use of nature – more specifically, realistically-portrayed flowers and plant forms. Louis Majorelle (1859–1926), who was one of the founding members of the Nancy School in 1901, took over his father's workshop, gave up ceramics and devoted himself exclusively to furniture. Blessed with an innate sense of form and incomparable technical virtuosity, he was artist, craftsman, designer and technician all in one. He is one of the great masters of Art Nouveau furniture. All the furniture he produced between 1898 and 1908, his most productive years, is dazzling in its elegance: for example, the *Orchids* desk in purplewood and rosewood, decorated with two corolla lamps designed by the Daum glassworks (1903). The work of Majorelle rivals that of the great 18th-century masters. He took over Bing's shop in Paris in 1904 and commissioned the architect Henri Sauvage to refurbish it. The firm of Majorelle Frères sold works by the Nancy School along with the stock of Bing's shop. After 1908, however, people were no longer willing to pay for quality, and the mechanization of workshops put an end to the production of extravagant items.

Hogweed upright drawing-room chair by Émile Gallé, 1902. The nature-inspired motifs are in walnut and kid leather.

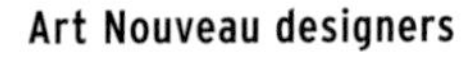

Art Nouveau designers

Émile Gallé (1846–1904). Gallé, a glassmaker, ceramicist and furniture designer, opened a marquetry and cabinet-making workshop in 1885, operating in parallel with his glassworks. In 1889, he showed two lines of carved wooden furniture at the World Fair in Paris. As he had done for his glassworks, he industrialized part of his workshop in 1894, mechanizing various stages of production. Thereafter the workshop was able to produce small-scale furniture such as tea tables, screens, stools, nests of tables and pedestal tables on a daily basis. Larger furniture, such as shelves, display cabinets and complete decorative schemes, did not appear until the late 1890s. Around 1900, the products were more sophisticated, Gallé having by then perfected his art. His *Dawn and Dusk* bed (1904) shows that just before he died he was on the threshold of a new, decisive phase in furniture design.

Hector Guimard (1867–1942). An architect and furniture designer, he was in Brussels when he met Victor Horta, who was finishing work on the Hôtel Tassel. As a result of this decisive meeting, Guimard developed a unique decorative language which he applied to his architectural work on the Castel Béranger apartment building in Paris (1894–8), integrating the interior decoration with the design of the communal and private areas. He used craftsmen, including the ceramicist Bigot, and art guilds.

Smoking-room bench seat (1897) by Hector Guimard, from the former collection of the pharmacist Albert Roy. It is made of profiled, carved jarrah wood and the metal is engraved. Made by Le Coeur & Co. Paris, Musée d'Orsay.

In 1898, his album *L'Art dans l'Habitation Moderne: le Castel Béranger, Œuvre de Hector Guimard* was published. This building was, in effect, the manifesto of his concept of total architecture, a place where crafts and art came together successfully. The Castel Béranger was followed by other projects: the Humbert-de-Romans concert hall (1898), the Hôtel Roy (1898), the entrances to the Paris metro and the Castel Henriette (1899–1900). Describing himself as an 'artist architect', Guimard designed every aspect of his buildings, including the furniture. His contribution to the 1900 Paris World Fair consisted mainly of decorative art designs. He fine-tuned the scheme he had used at the Castel Béranger, and for each interior he produced paintings, wallpapers, panelling, fireplaces, furniture, lights, vases, inkstands, door handles, and so on. Guimard based his art on three principles: logic, harmony and feeling.

Cabinet-makers. Nearly 50 cabinet-makers produced Art Nouveau furniture on a regular or occasional basis. For most designers, the annual salons were an unmissable opportunity to present their latest works. Critics rarely stayed away from the Champ-de-Mars or the Grand Palais, and their comments were decisive in furthering careers or causing them to sink without trace.

In 1898, the Parisian group – consisting of Alexandre Charpentier, Jean Dampt, Charles Plumet, Tony Selmersheim, Étienne Moreau-Nélaton and Félix Aubert – adopted the name 'L'Art dans Tout' and welcomed newcomers such as the architects Henri Sauvage and Louis Sorel, the sculptors Carl Angst and Jules Desbois, and the textile specialist Antoine Jorrand. They did not file any articles of association, but in their catalogue they set out ideas similar to those of Viollet-le-Duc: a rejection of the idea of style, a respect for materials and function, and a shared aim of designing a modern home.

Stylized forms: the Glasgow School

From 1895 to 1900, the Art Nouveau style spread across Europe. In the English-speaking countries and in Glasgow, Vienna and Munich, it moved away from historicism and the exuberant decorative excesses of Belgian and French Art Nouveau and towards a geometrization of forms and the invention of a modern style.

One event in particular had a decisive impact on the style in Glasgow. A new director of studies was appointed at the Glasgow School of Art in 1885: the remarkable Francis H Newbury. He encouraged his students to adopt a design philosophy that was not based on any prior stylistic assumptions, and he guided the school towards a free style in painting and decoration. The school was regarded as the most progressive in Europe until the Bauhaus opened its doors.

While serving his architectural apprenticeship, Charles Rennie Mackintosh (1868–1928) attended evening classes at the Glasgow School of Art from 1884, and during his eight years there he was awarded many prizes. He travelled in Italy before embarking on a prolific architectural career. He was entrusted with the designs for the Glasgow Herald Tribune building (1893–4), Queen Margaret's Medical College (1895) and the Martyrs' Public School (1895). The building of the Glasgow School of Art, constructed in two phases (1897–9 and 1907–9), is regarded as his masterpiece. In 1894, he developed an interest in illustration and designed the cover of a theatre programme, which incorporates one of the first examples of a stylized interpretation of nature. His friend and future partner Herbert Macnair was also interested in experimenting with new decorative forms. Mackintosh distinguished himself as a designer of interiors, furniture and decorative objects that always seemed just right, without superfluous ornament. His first furniture commission, for the Davidson family in Gladsmuir, dates from 1894–5. In 1895, he designed furniture for Guthrie & Wells, a well-known Glasgow factory; for them he produced furniture with simple lines that respected the natural qualities of the wood. That same year the organizers of the 'L'Œuvre Artistique' exhibition in Liège contacted Newbury to ask if he could send a selection of the school's work for their

Interior of the Hill House by Charles Rennie Mackintosh (1903). The whole scheme - armchairs in natural wood and high-backed upright chairs - is suffused with light. The hanging ceiling lamps with their geometric forms harmonize with the frieze running round the cornice. Hunterian Museum.

next event. Newbury selected the work of some students, including Mackintosh, and thus established the first link between the Glasgow Style and the Continent. Some of the decorative elements in the interiors designed by Mackintosh are the work of the MacDonald sisters: Margaret (1865–1933), who was Mackintosh's wife, and Frances (1874–1921), who married Mackintosh's partner Macnair.

The Four

The two couples formed a group known as the Glasgow Four. They explored every possible avenue in their search for an original style of decoration and ornamentation. Their stylized motifs lent themselves admirably to being executed in repoussé metal, plaster and stained glass. In 1896, they were invited to take part in the exhibition 'Arts and Crafts' in London, where they were excoriated by public and critics alike, who were shocked by the strange forms of their furniture. However, they were encouraged by the journalist Gleeson White in *The Studio* in 1896, and later by a major article in *Dekorative Kunst* in Darmstadt in 1898. Thereafter they were hailed as the leaders of a new artistic movement. The MacDonald sisters opened a studio in Glasgow where examples of the applied arts, embroidery, repoussé metalwork and light fittings were on display. It was much visited by literary and artistic figures, and the group members became celebrities of the Glasgow scene. In 1896, Mackintosh collaborated with George Walton in designing Miss

Cranston's tea rooms on Buchanan Street. Then, working alone, he designed the interior of her tea rooms on Argyle Street in 1897. Introducing a distinct difference of style, he gave an audacious line to all the elements: tables, umbrella stands, settees, small stools, armchairs and upright chairs. It was for this commission that he first came up with his well-known high-backed chair.

Between 1898 and 1904, Mackintosh's preference was for white painted furniture with touches of silver, mauve and green; in the light, the forms became almost ethereal. Rectangular high-backed, throne-like chairs; lights hanging from beaded chains; display cabinets with doors decorated on the inside – everything accentuated the fairytale setting. During this period he designed over 400 objects ranging from chairs to small forks. His art became increasingly abstract and geometric, although it lost none of its elegance or sophistication. He continued to work as an architect, travelling throughout Europe. The group exhibited at the 1900 Secessionist exhibition in Vienna, in the beautiful building constructed two years earlier by Josef Maria Olbrich (1867–1908). The Glasgow Four realized they had a lot in common with the Viennese designers. As an architect Mackintosh viewed the design of an interior as part of a whole, seeing all the elements, even the minor ones, as linked. He took part in an exhibition in Dresden in 1901 and another in Turin in 1902.

The Vienna Secession: 'Only that which is functional can ever be beautiful'

The influence of William Morris and his circle and of the Arts and Crafts societies was crucial in both Scotland and Vienna. Germany and Austria were strongly influenced by English designers, particularly Baillie Scott. In 1894, Otto Wagner (1841–1918) was appointed professor and head of the Special School of Architecture at the Vienna Academy of Fine Arts. As an innovative architect he contributed towards the dissemination of avant-garde ideas in the applied arts, and in 1895 he published his manifesto *Moderne Architektur*: 'Only that which is functional can ever be beautiful.'

The desire to change the eclectic, historicist style of the last third of the 19th century and the idea of parity between the craft-based arts and the 'noble' arts were two of the pillars on which the Vienna Secession was built. Felician von Myrbach took charge of the movement when it was founded in 1897, adopting a programme of progressive reforms. He gathered young, enthusiastic teachers round him: Josef Hoffmann for architecture, Koloman Moser and Alfred Roller for the decorative and applied arts, and Arthur Strasser for sculpture. The Vienna Secession made its first appearance on the applied arts scene with a major exhibition, 'The Jubilee of Arts and Crafts', held in spring 1898. 'Modern'

Sitting Machine armchair with adjustable back, made of mahogany-stained beech. Designed by Josef Hoffmann and produced by the company J & J Kohn, 1905. Saint-Étienne, Musée d'Art Moderne.

works were shown to the public for the first time. It was an instant success.

The interiors and furniture designed by Josef Maria Olbrich and Joseph Hoffmann became a particular focus of attention – Olbrich's for his elaborate forms and use of colours in the opulent tradition of Hans Makart (1840–84), and Hoffmann's for his advocacy of rationalism and simplicity. The Vienna Secession made a crucial contribution to the renewal of Viennese decorative art, becoming a mecca for the training of a new generation of Viennese artists, who would later collaborate in the Wiener Werkstätte (Vienna Workshops). On the façade of the Secession House (1897–8) in Vienna, built by Olbrich, the following motto can be read: *Der Zeit ihre Kunst, Der Kunst ihre Freiheit* (To each period its art, To art its freedom).

The Wiener Werkstätte

The year 1903 saw the opening of the Wiener Werkstätte (Vienna Workshops), run by Hoffmann and Moser, who also taught at the Vienna Secession. They were appointed artistic directors and secured the finan-

C Series set of glasses and carafe by Josef Hoffmann, 1912. Decorated with bronze-coloured geometric motifs on opaque glass, they were produced by the company J & L Lobmeyr, Vienna.

cial backing of Fritz Waerndorfer, who had just returned from a trip to England, where he had met Mackintosh. This financial support allowed them to make their scheme a reality. At their meeting, Waerndorfer and Mackintosh had come up with the idea of establishing production workshops run along cooperative lines. Although these were dedicated to craftsmanship, they were equipped with modern machines which allowed products to be closely monitored. The objectives of the undertaking were to 'promote the economic interests of its members by providing teaching and training in the arts and crafts; by manufacturing all types of craft-made objects, produced in accordance with the artistic concepts of the members of the association; and by the founding of workshops and the sale of the manufactured objects'. Hoffmann and Moser set up workshops for metalworking and gold and silver work, workshops for bookbinding, leatherwork and lacquer work, a workshop for cabinet-making and a drawing office in a former factory. They made arrangements for their architectural practices to work in conjunction with the workshops and sought out qualified craftsmen – bookbinders, goldsmiths, carpenters and painters. Their aim was to look with fresh eyes at all objects used in everyday life and in the decorative arts. With a properly equipped workplace, healthy working conditions and modern plumbing, the Wiener Werkstätte became a talking point. In October 1904, when their first exhibition opened in Berlin, the magazine *Deutsche Kunst und Dekoration* ('German Art and Decoration') devoted 46 large-format, lavishly illustrated pages to the event.

In October 1905, the Werkstätte held an exhibition in the new rooms of the Mietke Gallery, which had been fitted out by Hoffmann; they showed sculptures by Richard Luksch, jewellery by Carl Otto Czeschka and a maquette of the Palais Stoclet in Brussels. The Palais Stoclet (1905–11), designed by Hoffmann, was the largest project completed by the Wiener Werkstätte. Gustav Klimt decorated the dining-room with large mosaics, Johann Soulek designed furniture, and Hoffmann designed furniture, glass objects and silverware. This was a total work of art: working together, the members of the workshops created the whole interior right down to the smallest detail. In 1910, Eduard Wimmer developed a fashion department within the Werkstätte which employed up to 80 people. Other completed schemes included the Pukersdorf sanatorium, the sales premises of the Seilerstätte State Printworks, and private houses commissioned from Hoffmann. From 1906 onwards, the Wiener Werkstätte took part in international exhibitions. The 'Emperor Casket' by Czeschka was displayed at the Imperial Royal Austria Exhibition in London in 1906 and was the focal point of the show.

The products of the Wiener Keramik factory, founded in 1906 by Michael Powolny and Berthold Löffler, were sold by the Werkstätte from 1907 onwards. Financial difficulties led to Moser's resignation, and other collaborators came on board. The Fledermaus Cabaret, a theatre incorporating artwork by many members of the Wiener Werkstätte, opened in 1907. That same year, a Wiener Werkstätte shop was opened in Vienna, offering luxury products with modern, restrained, geometric forms. In 1912, Otto Wagner built a block of artists' studios and rented some of them to the Wiener Werkstätte. He also made materials available to the artists free of charge, allowing them to pursue their experiments. The resulting craft objects were featured in the shop's sales catalogue and royalties were paid to the artists. In 1914, following the declaration of war, Fritz Waerndorfer had to leave Austria for the USA. The company was now in difficulties. In 1915, Hoffmann brought Dagobert Peche (1887–1923) to the Wiener Werkstätte. Until his death he would be one of its most important collaborators.

A community experiment: the Darmstadt Artists' Colony (1899–1914)

In 1892, the Grand Duchy of Hesse passed into the hands of the 23-year-old Ernst Ludwig von Hessen und bei Rhein. His mother had died prematurely, and his grandmother, Queen Victoria, had taken a personal interest in his education. He paid many visits to the British court. He probably knew some members of the Arts and Crafts movement, and his activity as a patron began with a commission to Baillie Scott for the furnishing of his wife's two drawing-rooms (1897). Eager for Hesse to prosper and for art to flourish there, his efforts in this regard brought him

Poster for the Darmstadt Artists' Colony exhibition of May–October 1901. The graphic design is by Josef Maria Olbrich.

considerable distinction. With its industrial dynamism, its building boom, its growing economy and its efficient, mechanized furniture factories, Darmstadt was the ideal place for the development of Jugendstil (Art Nouveau) ideas. In January 1890, the publisher Alexander Koch brought out *Deutsche Kunst und Dekoration*, the first German monthly magazine devoted to the decorative arts. The trend was towards the development of an art that was national, autonomous and modern.

With the help of the architect Josef Maria Olbrich, who proved to be an ideal collaborator, Grand Duke Ernst Ludwig founded the Künstlerkolonie (Artists' Colony) in 1899. He brought together seven artists: Olbrich himself, who had a profound influence on Darmstadt

Jugendstil and can be seen as the heir to Otto Wagner, the famous architect of the Vienna Secession; the painter and graphic artist Peter Behrens, whose work showed how painting could be successfully applied to the decorative arts; the painter Hans Christiansen; the architect and decorator Patriz Huber; the sculptor and medal-maker Rudolf Bosselt; the decorative painter Paul Burck; and the sculptor Ludwig Habisch. The seven members of the Artists' Colony were contracted to work for a period of three years in exchange for a basic income. A committee was set up by the Grand Duke to encourage cooperation between artists, craftsmen and industry and thereby generate extra income.

The concept of total art

The Colony wanted to create a synthesis between art and production in the spirit of the Arts and Crafts movement. It aimed to achieve an aesthetic reform of all forms of art, and it invented a type of architecture and a means of producing consumer goods that were linked to modernity. It was not interested in creating fashionable artistic movements, but rather a new form of contemporary art with a direct relation to everyday life.

Not just painting and sculpture but all fields of craftsmanship were involved; however, architecture and design were the dominant areas. The artists in the Colony saw decoration as a total, homogeneous activity, from the shape of the roof right down to a small spoon. At a major exhibition in 1901 entitled 'A Testimony to German Art' (in preparation since 1899), Olbrich presented plans for a whole district of the town, based on a system of octagonal axes. This was to include the workshop building, an exhibition centre called the Ernst-Ludwig-Haus, five artists' houses, an open-air theatre and an observation tower. Olbrich himself was the architect for this unique development. The artists' houses were custom-built; only Behrens, who was also an architect, built his own house. The Colony artists took care of the interior decoration and the furniture.

This was the Artists' Colony's concept of a total work of art: unity between art and life, between artist and craftsman, and between house and furniture. This aesthetic angle on life was a matter of course for Peter Behrens, who wrote to Ernst Ludwig in 1901: 'In this way beauty becomes for us the quintessence of the supreme power, and a new cult is born to serve it. We want to erect a house to it, an abode where art is formally displayed, and devote our lives to it there'. ('Ein Dokument Deutscher Kunst', *Festschrift*, Munich)

The end of the Artists' Colony

The Colony's originality lay in this tension between aestheticism and economics, but the 1901 exhibition was a financial failure, leading to serious dissension within the group. Behrens left the Colony in 1903, although Olbrich remained in it until his death in 1908. After establishing the Künstlerkolonie, the Grand Duke set up a ceramics factory managed by Jakob Schwarvogel (1904–6), a factory for decorative glass run by Emil Schneckendorf (1906), a training workshop for the applied arts (1907) and an 'Ernst Ludwig' printworks.

Financial difficulties increased. Then Emanuel Josef Margold joined the Colony. Trained as one of Hoffmann's top pupils, he introduced a 'Viennese' spirit into he group, decorating his interiors, tapestries, carpets and ceramic objects with borders featuring stylized flowers. In 1912, he created a kind of 'corporate design' for the Hermann Bahlsen biscuit factory, using the same motif in various ways in the decoration of shops, packaging and displays. In 1911, a third architect, Edmond Körner, joined Margold and Bahlsen. In 1914, preparations were made for the final exhibition of the Darmstadt Artists' Colony, showing works that were stylistically located somewhere between Jugendstil and Art Deco. That same year the Deutscher Werkbund (German Working Alliance) mounted a major exhibition in Cologne. When war broke out in August 1914, the Darmstadt Künstlerkolonie was disbanded. After the war, a different group took on the leading role: the Bauhaus, founded in Weimar in 1919.

The German Secession

From 1892 onwards, Munich was home to the modern movement. This consisted of secessionist artists who were united in their rejection of official art and who aspired to change the everyday environment. In 1897, Hermann Obrist, Bernhard Pankok, Peter Behrens, Julius Scharvogel, Ludwig Habich, Richard Riemerschmid and Bruno Paul founded the Vereinigte Werkstätten für Kunst im Handwerk (United Workshops for Art in Craft), which had its roots in Viennese Art Nouveau and the Arts and Crafts movement. They organized an exhibition called 'Art in Craftsmanship' (1901) which was purely decorative in inspiration and devoid of any social reference.

The Deutscher Werkbund and design

In 1896, Herman Muthesius went to London on secondment to the German Embassy. When he came back in 1903, he wrote *Stilarchitektur und Baukunst* ('Style-Architecture and Building-Art'), an account of architecture and the decorative and industrial arts. In this he describes the role of the machine, which he calls the 'producer' of standardized artistic products. After a close study of Ashbee's experimental work, Muthesius was the first to formulate the principles of industrial aesthetics: opposition to the Art Nouveau style, criticism of 'ornamental artists', and reorganization of the management of the schools of applied art (the 'Deutsche Werkstätten'). He won over young architects such as Bruno Paul and Behrens. Picking up where Darmstadt left off, a trend towards

Oak and leather chair by Richard Riemerschmid, c.1900. Bethnal Green Museum collection.

simplicity became widespread, and numerous design studies were produced for furniture intended for the working classes. At Hellerau, Behrens and Karl Schmidt designed standard items of furniture, which Muthesius described as 'mechanized furniture'. Heinrich Tessenow and Bruno Taut followed on from him, giving serious thought to the design of working-class housing with a view to improving the living conditions of the working classes. These were the core ideas of Muthesius, the spiritual father of the Werkbund: to create a middle-class culture for the industrial age and to bring about a continual rise in status of the working class so that it could participate more and more in this new middle-class culture.

The Werkbund was a German association of artists, architects, craftsmen and industrialists founded in Munich in 1907 with the aim of 'ennobling the profession of art through the joint action of art, industry and the crafts'. This association was soon playing a leading cultural role, having a decisive influence on the formal development of housing, furniture and utilitarian objects. Enhancing the prestige of work and improving quality – either through the alliance between art and industry or through the use of craftsmanship within a modern context – promoted social cohesion, which was one of the association's guiding principles. Muthesius, Behrens, Gropius, Van de Velde, Hoffmann, Taut, Riemerschmid and Adolf Meyer were among the founding members.

The Werkbund was not a German version of the Arts and Crafts movement. While it shared many of its principles, the group's commitment to collaborate with industry rather than oppose it was a major difference; it attempted to reconcile inventor and maker, as well as invention and form. In his inaugural speech, Fritz Schumacher urged the Werkbund to focus on the form given to machine-made products. In doing this, he showed an innovative interest in design and lent the movement an aura of modernity from the very beginning. This new way of looking at things was responsible for the association's success and for the interest shown in it by industrialists. In 1907, the company AEG invited Behrens to review the design of their whole business. This aesthetic decision also served political goals: the industrialists wanted to get the workers on their side by ensuring good working conditions in the form of well-lit and well-ventilated factories, thereby weakening the

Peter Behrens

ELEKTRISCHE TEE- UND WASSERKESSEL

NACH ENTWÜRFEN VON PROF. PETER BEHRENS

Messing vernickelt, streifenartig gehämmert runde Form

PL Nr	Inhalt ca. l	Gewicht ca. kg	Preis Mk.
3581	0,75	0,75	19,–
3591	1,25	1,0	22,–
3601	1,75	1,1	24,–

Kupfer streifenartig gehämmert runde Form

PL Nr	Inhalt ca. l	Gewicht ca. kg	Preis Mk.
3584	0,75	0,75	20,–
3594	1,25	1,0	24,–
3604	1,75	1,1	26,–

Messing streifenartig gehämmert runde Form

PL Nr	Inhalt ca. l	Gewicht ca. kg	Preis Mk.
3582	0,75	0,75	19,–
3592	1,25	1,0	24,–
3602	1,75	1,1	25,–

ALLGEMEINE ELEKTRICITÄTS-GESELLSCHAFT

ABT. HEIZAPPARATE

AEG (Allgemeine Elektricitäts-Gesellschaft) made Peter Behrens responsible for the total concept of the company's brand and design image. He designed this page from their catalogue showing electric kettles (graphic design, page layout, typography), 1909.

Peter Behrens (1868–1940), a German architect, painter and industrial designer, was a co-founder of the Munich Secession in 1890. Influenced by the Arts and Crafts movement, he belonged to a generation of artists who moved away from the plastic arts and directed their talents towards industrial production. Together with Hermann Obrist and Bernard Pankok he was involved in setting up the Vereinigte Werkstätten für Kunst im Handwerk in Munich, with the aim of giving everyday life some sort of artistic unity. Having been invited to join the Darmstadt Artists' Colony in 1899, he built his own house there and helped to prepare the German contribution for the Turin exhibition of 1902. He became head of the School of Decorative Arts in Düsseldorf, which he turned into a major German centre of cultural reform. As one of the founding members of the Werkbund in 1907, he was given a special commission by AEG. He was asked to rethink the company's image in its entirety: from its factory buildings, workers' estates and shops to the typeface design for its publications, advertisements and logos. He was also required to design such items as arc lamps and electric kettles. As a consequence of this unprecedented commission, Behrens can be regarded as the first industrial designer. He designed for other companies – a sewing machine for Pfaff in 1910, for example. Behrens thought of the factory as a 'cathedral of work'. In his booklet *Feste des Lebens und der Kunst* ('Celebration of Life and Art'), he expressed his desire to overcome 'commercialism' by means of beauty. In 1912, he was approached by a trade union to design furniture suitable for ordinary people. In 1926, he joined the group of Berlin architects known as Der Ring, founded by Erich Mendelsohn and Hans Poelzig, and from 1922 to 1936 he was Professor of Architecture at the Vienna Academy, then at the Academy of Fine Arts in Berlin. Major architects passed through his practice in Berlin on training placements: Gropius, Le Corbusier, Ludwig Mies van der Rohe and Paul Thiersch.

basis for social democracy. The attractive, sturdy items produced by the German factories led to the expectation that they would do well on the international market. Friedrich Naumann, the ideologist of the Deutscher Werkbund, issued a reminder that products had to be of a high quality. But, for Muthesius, the products also had to have a style that was identifiably German. Those involved in the Werkbund wanted Germany to be a powerful nation. Art put itself at the service of industry; it was not a luxury but an economic force. The Werkbund worked in the field of industrial design: Behrens designed arc lights, Neumann designed cars, and Gropius designed railway engines and sleeping cars.

For or against standardization

The Werkbund organized a major exhibition in Cologne in 1914, when there was much debate about the apparent contradiction between standardization and free art. In advocating standardization in his lecture on 'The Werkbund's Work for the Future', Muthesius wanted to guide the movement in the direction of industrial design and replace the 'extraordinary' by the 'ordinary'. Van de Velde, representing free art, argued vehemently against this, championing the artist's work, his freedom to decide and his refusal to submit to the rigid standards of the machine.

World War I and defeat in 1918 shattered all hopes of German economic supremacy; both factions in the Werkbund accused industry of being responsible for the war. The Werkbund then turned its back on industry, returning to its point of departure: William Morris and the Arts and Crafts movement. Functional form, the aesthetics of the machine and industrial design were no longer at issue; the talk was of craft work and artistic quality. Gropius decided to take the arts back to the ideals of craftsmanship and founded the Bauhaus in 1919.

Following the conference that the Deutscher Werkbund held in Vienna, an Austrian Werkbund was set up in 1913. The German model was also followed in other countries, such as Switzerland (1913). The aim of the Swiss Werkbund was to disseminate modern culture by means of industrial and craft production.

The invention of design (1914-39)

· Art and design: the avant-garde in search of a three-dimensional visual space
· Functionalism and prefabrication
· Paris: an inventory
· Industrial design in the USA
· Scandinavian design, organic design
· Biomorphism
· Power and design

With the development of European avant-garde movements, designers began to think radically about the image and function of objects. In Germany, the Bauhaus was the first to come up with a definition of design based on the study and theory of production methods and on its commercial ambitions. Meanwhile, France sought to bridge the gap between the decorative and industrial arts which had been so apparent at the Paris International Exhibition in 1925. The USA led the way in making design a beacon of a great industrial nation. After the upheavals caused by the Wall Street crash, designers became key players in the consumer society. In 1937, the year of the Paris World Fair, progress once again became the power symbol of developing nations.

Art and design: the avant-garde in search of a three-dimensional visual space

Around 1910, the avant-garde was asking fundamental questions about design: is there a difference between the production of aesthetic objects and the production of industrial objects? What are the relations between three-dimensional visual space and the space we live in? New forms of expression developed simultaneously in various avant-garde movements: the Cubist painters in France, the Futurists in Italy, the Constructivists in Russia, the artists centred on the De Stijl group in the Netherlands, and the Dada movement.

Fernand Léger: a language of the modern world

The war left its mark on artists. Fernand Léger served on the front line as a stretcher-bearer, an experience that altered his vision of reality. 'It was during the war that my feet really sank into the earth,' he said. From then on his works depicted subjects drawn from everyday life. As early as 1913, he had been the first to express, in a systematic way, the response of avant-garde circles to the renewal of forms and rhythms triggered by the industrial world. Advertising with its jazzy graphic approach and the intrinsic beauty of machines and industrial objects were seen as sources of excitement. Léger recognized engineers and technicians as inventors of forms. His meeting with the architect Le Corbusier in 1920 led him to explore the relationship between art and architecture in monumental paintings.

Previous page: German pavilion at the 1929 International Exhibition in Barcelona. Architecture and furniture by Ludwig Mies van der Rohe: onyx partition, Barcelona chairs. Sculpture by Georg Kolbe.

'A car ... more beautiful than the Winged Victory of Samothrace'

From 1909, when an article serving as its manifesto was published in *Le Figaro*, the artistic experiments of the Futurist group were centred on the poet, painter and polemicist Filippo Tommaso Marinetti (1876–1944): 'Let's destroy museums – those cemeteries; a work has to be aggressive ... A roaring car hurtling along like a machine-gun is more

Painted wooden tray with an abstract pattern and overlapping reliefs. Giacomo Balla, 1920. Balla turned his home, 'La Casa Futurista', into a living museum of Futurism.

beautiful than the Winged Victory of Samothrace'. Futurist art aimed to be committed, socialist and humanitarian, combining nationalism and a love of progress. The Futurists championed an ideal vision of the machine and drew inspiration from scientific images (as found in motion sequence photography or industrial drawings). Their works expressed the speed which – as they saw it – characterized the industrial age, the age of the car and the train. They explored many different fields: poetry, the visual arts, photography, fashion, cinema and the decorative arts. In favour of a total renewal of urban art, they sought to exert their influence primarily on architecture and town planning. However, they also operated in the areas of interior design and furniture, as is demonstrated in particular by the works of Giacomo Balla (1871–1958), Fillia (the pseudonym of Luigi Enrico Colombo, 1904–36) and Nicola Diulgheroff (1901–82). Antonio Sant'Elia (1888–1916) published *Città Nuova* (1913, 'New City') and the 'Manifesto of Futurist Architecture' (1914). This utopian vision of the city represented a challenge, especially in the economic and social context of Italy at the time.

Together with Fortunato Depero, Balla published the first Futurist manifesto on the applied arts: 'The Futurist Reconstruction of the Universe' (1915). In 1920, he opened 'La Casa Futurista' in his private house, decorated entirely in accordance with his own instructions. Fillia founded the Turin Futurist movement and the Futurist artists' guilds, which adopted an uncompromisingly revolutionary and proletarian stance. He took part in Futurist reunions from 1922 onwards, experimenting in the fields of painting, decoration – *Ambiente Novatore* (Turin 1927, 'Innovative Decoration') – furnishing and ceramics (with Tullio d'Albisola).

Russian Constructivism: 'art in life'

The advent of the Soviet regime in 1917 gave a boost to avant-garde movements, especially in the field of the applied arts. However, the Tenth

Red and blue Neo-Plastic armchair by Gerrit Rietveld, 1918. It uses spatial overlapping and colours in the spirit of the painter Piet Mondrian.

National State Exhibition, 'Non-Objective Creativity and Suprematism' (1919), revealed certain tensions. There were two opposing tendencies: supporters of the Suprematism of Kasimir Malevich (1878–1935) and advocates of the Constructivism of Vladimir Tatlin (1885–1953), including Alexander Rodchenko (1891–1956) and Eliezir Markovich, known as El Lissitzky (1890–1941).

Suprematism, dominated by the painter Kasimir Malevich, aimed to subject the very materiality and structure of objects to the abstract experiments which were already being carried out in the visual arts. Malevich applied this approach to ceramics and put forward a design for an urban environment based on the same constructive logic. Piet Mondrian (1872–1944) said that Malevich's easel paintings were like photographic plates waiting to be developed, sensitive to the wall space of each individual room. After the major retrospective of his work held in Moscow in 1919, Malevich devoted himself to teaching and theoretical experiments. He left Moscow for Vitebsk, where he became head of the school of art, which he named Uno-Vis (College of New Art). He worked out a teaching method based on idealism. The brothers Naum Gabo (the pseudonym of Naum Neemia Pevsner, 1890–1977) and Antoine Pevsner (1884–1962) shared Malevich's ideas: that the aim of art is to make concrete the vision that man has of the world.

For their part, the Constructivists made their presence felt with a new programme: 'art in life'. They laid the foundations of the productivist theory (or the sublimation of unalienated labour) in 1918, within the Izo (plastic arts section) of the Narkompros (People's Commissariat of

Enlightenment). The artists Vladimir Tatlin, Alexander Rodchenko, his wife Varvara Stepanova (1894–1958), Liubov Popova (1889–1924) and Alexander Vesnin (1883–1959) devoted themselves primarily to design, thus bringing together the artist and the engineer. Art and industry were both subject to the same economic and technical requirements. Rodchenko, a friend of the poet Mayakovsky, ran experimental workshops with El Lissitzky: the Vkhutemas-Vhutein (Higher State Artistic-Technical Workshops). Founded by decree in various regions of the USSR in 1920, these were a response to 'the need to train highly qualified artist-engineers for industry'. Unlike the Bauhaus, the workshops taught architecture alongside subjects relating to the perception of forms, volumes and colours. The teachers – architects, painters, sculptors, designers and graphic designers – were the leaders of the artistic avant-garde. Rodchenko designed multi-purpose furniture. The October Revolution of 1917 allowed Constructivism to cross the threshold separating utopia from reality. In 1921, Rodchenko declared that the production of objects was the sole purpose of Constructivist creativity and defined an early form of functionalism. In 1923, the first issue of the Constructivists' journal *Lef* stated: 'the material formation of the object will replace its aesthetic make-up', thus defining a role for the Constructivists in the economic life of the country. Tatlin, the founder of the Constructivist movement, took a job in a metallurgical factory; Popova and Stepanova collaborated with textile factories; and Rodchenko and Mayakovsky worked together on books and posters.

De Stijl, an international union of life, art and culture

The same revolution in industrial aesthetics also took place in the Netherlands, which had remained untouched by the war. The painters Piet Mondrian, Bart van der Leck (1876–1958) and Theo van Doesburg (1883–1931) produced the avant-garde magazine *De Stijl*, the first issue of which appeared in October 1917. The magazine was used by Mondrian to publish his theories. Inspired by the research of Dr Mathieu Schoenmaekers, who had just published his *New World Vision* (1915), Mondrian concluded that art can be just as exact as mathematics in expressing the fundamental characteristics of the universe. In his 1920 essay on Neo-Plasticism (a theory that art should concern itself only with absolute qualities, developed by Mondrian and van Doesburg), Mondrian wrote: 'The decorative arts disappear in Neo-Plasticism'. Everyday life was to be the arena for the synthesis of the plastic arts. The magazine *De Stijl* was published until 1931.

Several painters and architects were to join the group: Vilmos Huszár (1884–1960), Georges Vantongerloo (1886–1965), Jacobus Johannes Pieter Oud (1890–1963) and Robert Van't Hoff – who had come back from the USA, where he had met Frank Lloyd Wright. They all shared the belief that 'works of art must be created by the collective spirit and must try to bring together the various disciplines. They must be present in everyday life, and everyday life must be seen from now on as a synthesis of the plastic arts' (Michel Seuphor, *Le Style et le Cri*, 1933). Van Doesburg advocated the 'exclusive use of right angles positioned horizontally and vertically, the three primary colours and the non-colours white, black and grey.'

The movement found its expression in fine art, furniture and architecture, promoting a vision of the world based on abstraction. Among the works which implemented their somewhat extreme commitment were: the De Unie café (1924) by Oud, the 'Neo-Plastic' armchair (1918) by Gerrit Rietveld (1888–1964) – the first design object, according to Van Doesburg – and the interior planning of the Villa Schröder in Utrecht (1924), an example of total art.

Dada and design

In his machine paintings, Francis Picabia was the first to introduce illogical titles which undermined appearances by the use of irony. Marcel Duchamp subverted reality in the same way with his 'ready-mades'. He entitled a work *Snow Shovel* – which is exactly what it was, no more and no less. He challenged the status of art and created anti-art, presenting the bottle-rack and the urinal as objects worthy of veneration, arro-

Bicycle Wheel, Marcel Duchamp, 1913. Ready-made. Assemblage of bicycle wheel and stool. Paris, Musée d'Art Moderne.

gantly criticizing the history of forms and playing about with objects. The ready-made – the ultimate trompe l'œil – marked the transition from the reproduction of nature to nature itself. The boldness of Picabia and Duchamp influenced the American artists Arthur Dove and Morton Schamberg, who also used machines to subvert meanings.

Functionalism and prefabrication

In contrast to England with its Arts and Crafts movement, Europe and the USA became fascinated with technology, which it began to glorify. The new materials developed by scientists – such as steel, reinforced concrete, aluminium and linoleum – brought with them a new structural logic. The term 'functionalism' sums up the ideas of those who designed

Teapot, silvered bronze and ebony, Marianne Brandt, 1924. Produced by the metal workshop of the Bauhaus in Weimar.

the everyday items produced by industry, at the Bauhaus or elsewhere. Producing an object that functions well implies studying its function, in other words the way it is used. Form follows function. Function was defined through studies which clarified the different aspects of use and through experiments that took account of scientific and technical knowledge, the requirements of industrial manufacture and the possibilities afforded by materials. In 1910, Peter Behrens drew up 'a memorandum on the industrial prefabrication of houses on a unified artistic basis'. This idea of prefabrication was taken up in the USA. Walter Gropius also supported it, putting it into practice with the Fagus factory at Alfeld-an-der-Leine, and then in 1914 with his office building at the Werkbund exhibition in Cologne (working in both cases with Adolf Meyer).

The Bauhaus: the Weimar period (1919-24)

The Bauhaus represents a crucial coming together of ideas and achievements related to Germany's economic and political situation. The school was formed by merging the Academy of Fine Arts and the School of Arts and Crafts in Weimar; the latter had previously been run by Henry van de Velde. Gropius presided over the new establishment, officially named the Staatliches Bauhaus. On 1 April 1919, he moved into buildings designed by Van de Velde. The school broke with middle-class conservative culture, becoming a centre of experimentation, producing rivalry between avant-garde artistic movements and creating a movement for cultural reform. Gropius favoured ties between art and industry. In the Bauhaus manifesto (published in April 1919), the first stated objective was to prevent the arts from being compartmentalized and to encourage craftsmen, painters and sculptors to combine their talents. The second objective was to raise the status of craftsmanship to the same level as that of fine art. 'There is no difference between the artist and the craftsman. The artist is a glorified craftsman.' Gropius brought in painters to carry out the teaching duties of the Bauhaus – he had learnt from the history of art that it is painting that brings about the renewal of forms. The third objective was to interest industrialists in the project, develop a production policy and achieve economic independence.

A new way of teaching

Gropius favoured courses taught by painters: Lyonel Feininger, Johannes Itten, Gerhard Marcks, Georg Muche, Oskar Schlemmer, Paul Klee, Lothar Schreyer, Wassily Kandinsky and László Moholy-Nagy. As he saw it, the plastic arts and the applied arts were inseparable activities. Johannes Itten (1888–1967) was one of the first three *Formmeister* (masters of form) at the Bauhaus; the others were Marcks and Feininger. Itten had been running a private school in Vienna since 1916, using anti-conformist teaching methods. In 1919, he was taken on at the Bauhaus by Gropius, who had met him a year earlier. He reformed the structure of the teaching system and introduced the preliminary course, or *Vorkurs*, an experimental period of one semester, at the end of which students chose a workshop where they studied for three years. He was also in charge of the workshops for stone carving, metalwork and painting on glass. Reflecting the commitment to a synthesis of the arts, each workshop was run by two masters, a *Formmeister* and a *Lehrmeister* (master of technique). Itten was a mystic, a disciple of Mazdaism. He made his students follow a programme of physical and mental exercises; he shaved his head and wore long, flowing robes. Gropius made him his leading collaborator at the Bauhaus. His teaching was based on a detailed study of nature, the making of three-dimensional compositions using a variety of materials, and the analysis of historical works of art.

Walter Gropius

The experimental kitchen of the *Haus am Horn*, 1923. Gropius's objective for this model home was to make it as convenient to use as possible. Marcel Breuer's kitchen, for example, is simple and rational and can also be used as a dining-room. Designed by Muche, this experimental house was exhibited as a possible prototype for cheap, mass-produced accommodation. It also served as a showcase for the talents of the Bauhaus workshops.

Born in 1883, Gropius began his career as an architect in 1907, working for Peter Behrens in Berlin. He opened his own practice in 1911. In the spirit of the Werkbund, he developed ideas about the functionalism of habitable space and designed a sleeping-car compartment for the Deutsche Reichsbahn (German Imperial Railways) in 1914. He organized the merger of the Academy of Fine Arts and the School of Arts and Crafts in Weimar into a single establishment, the Staatliches Bauhaus; this took place in 1919, with Gropius himself becoming its director. In the Bauhaus manifesto of 1919, he set out his ambitions for the school: 'Let us therefore create a new guild of craftsmen without the class-distinctions that raise an arrogant barrier between craftsmen and artists! ... It will combine architecture, sculpture and painting in a single form.' He dreamt of an ideal artistic community. In 1922, he erected a monument consisting of geometric concrete forms in the cemetery of Weimar, in commemoration of the striking workers who were killed in the Kapp Putsch of 1920. In 1925, the school moved to Dessau.

Gropius resigned from the Bauhaus in 1928 in order to devote himself to architecture. He went to the USA, where he studied the industrialization of housing and met Richard Neutra. In 1930, he organized the first German contribution to the Salon des Artistes Décorateurs in Paris. The Depression of the 1930s led to a slowdown in commissions. Gropius decided to resign from the Werkbund. In 1934, he left Germany for London, where he went into partnership with Maxwell Fry and took over responsibility for design at Isokon Ltd, an architectural and interior design practice. He left for the USA in 1937, and ran the Graduate School of Design at Harvard until 1952. From 1937 to 1944 he worked with Marcel Breuer, and in 1946 he opened his last practice in the USA. He died in 1969.

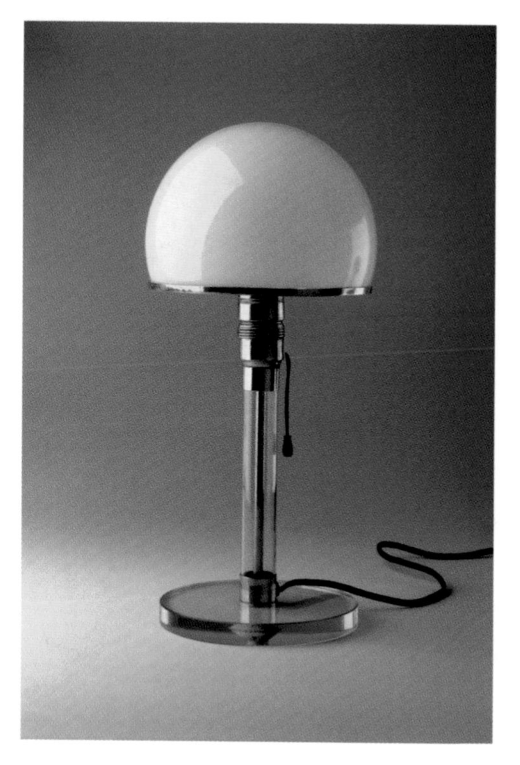

Metal and glass table lamp with an opaline shade, Wilhelm Wagenfeld and Jakob Jucker, 1923-4. Made by the metal workshop of the Weimar Bauhaus.

Unable to agree with the production-oriented direction the school was taking, he resigned in 1923.

Gerhard Marcks (1889–1981), a sculptor and a member of the Werkbund, designed a series of animal figures which were manufactured by the Schwatzburg porcelain factory. His experience of collaborating with industry was invaluable to the Bauhaus. He was appointed *Formmeister* of the pottery workshop. The German-American painter Lyonel Feininger (1871–1956) remained at the school until it was closed down. He was *Formmeister* of the print workshop, and made the wood engraving that illustrated the manifesto (1919).

Between 1920 and 1922, Gropius appointed four new *Formmeister*, all of them painters. Oskar Schlemmer (1888–1943) was responsible for the mural painting workshop, and then from 1923 for the stage workshop. Paul Klee (1879–1940) ran the design theory course, where he set out his theory about elementary forms. Wassily Kandinsky (1866–1944) provided an introduction-to-design course based on the study of colour and a course on analytical drawing. He championed the concept of the *Gesamtkunstwork* (total work of art), based on a synthesis of the arts. László Moholy-Nagy (1895–1946) succeeded Itten in 1923. He wore the same clothes as industrial workers – machines were a kind of fetish for him. He refocused the preliminary course on introducing pupils to basic techniques and materials and the rational use of these. In the metal workshop he tackled practical problems – the making of lamps, teapots and tea balls – and encouraged the use of steel. The teaching at the Bauhaus covered many other disciplines: stained glass, pottery, metal, weaving, theatre, fresco, architecture and typography.

In 1923, the Thuringian government asked the Bauhaus to exhibit work done at the school. Every department sprang into action. The theme chosen for the exhibition was 'Art and Technics: a New Unity'. An experimental house known as the *Haus am Horn* was constructed, a prototype for a cheap, mass-produced building using new materials. The furnishings and fittings of the house were all made in the workshops: carpets, radiators, tiles, and so on. The lamps were designed by Moholy-Nagy and made in the metal workshop, and most of the furniture was designed by Marcel Breuer, then a pupil at the school. He also designed

The *Wassily* armchair (thus named in homage to Kandinsky), tubular steel and leather, Marcel Breuer, 1925. Breuer, the carpentry workshop *Formmeister* at the Bauhaus, designed tubular steel furniture inspired by the handlebars of his bicycle.

an innovative, functional kitchen. The exhibition attracted 15,000 visitors. Between 1919 and 1924, a total of 526 students were trained at the school.

Josef Albers (1888–1976), a former student, became Moholy-Nagy's assistant. Taken on as a master of the glass-painting workshop, he gave instruction in the use of materials. During this period of political instability, the Bauhaus was threatened by the right, which was gaining ground in Weimar. Grants were cut, and in 1924 the director and the council of masters, supported by the students, decided to close down the Bauhaus. Many cities offered it a new home. The whole school eventually transferred to Dessau in the spring of 1925.

The Bauhaus: the Dessau-Berlin period (1925-33)

The Bauhaus opened in Dessau in 1925, in more favourable economic circumstances. The mayor of Dessau, Fritz Hesse, saw its installation there as an artistic boost to the town. The local council commissioned Gropius's practice to construct the new building for the school, as well as seven houses with studios for the teachers and a wing containing 28 flats for the students. The building was officially opened in December 1926. The interior was decorated and furnished by the students under Gropius's direction. Most of the teachers moved to Dessau along with the school: Feininger, Kandinsky, Moholy-Nagy, Muche and Schlemmer. A new generation that had trained in Weimar joined the teaching staff: Albers, Herbert Bayer (put in charge of the printing department, where he introduced advertising techniques), Breuer, Hinnerk Scheper and Joost Schmidt. The Bauhaus set out its principles clearly and consolidated its position as an avant-garde school. The pupils carried out experimental work, especially in the fields of architecture and decoration, and devoted themselves to the study of models for industrial and craft-based production. The output of the Bauhaus was relevant to everyday life: tubular steel furniture, modern textiles, crockery, lamps and modern typography.

The cabinet-making and metal workshops merged to form a new department run by Marcel Breuer (1902–81). He made the *Wassily* tubular steel armchair (1925), which was innovative in both technique and form, but not yet ready for mass production. He also designed a cantilevered chair produced by Thonet. He was the first to use steel, a product not found in traditional craft work. Research into metal furniture was also carried out by Mart Stam and Mies van der Rohe, with a view to industrial production. In 1927, Gropius set up an architecture department, headed by the Swiss architect Hannes Meyer (1889–1954).

The final flourish

Gropius left the Bauhaus at Dessau in 1928. Meyer took over from him, guiding the school towards a rationalist ethos totally bereft of romanticism: it had to meet the needs of the people. Meyer urged the workshops to give priority to producing objects and designs that could be sold to industry and to making inexpensive essential items such as furniture. The Bauhaus set up a company to exploit commercially the models designed in its workshops. Projects were conceived in collaboration with industrialists or manufacturers with a view to providing technical follow-up. Commercial negotiations were conducted by the Bauhaus itself, and the revenue was divided up between school, business partner and designer.

Teachers at the Bauhaus in Dessau, 1926. From left to right: Albers, Scheper, Muche, Moholy-Nagy, Bayer, Schmidt, Gropius, Breuer, Kandinsky, Klee, Feininger, Stölzl, Schlemmer.

Display of chairs by Marcel Breuer, tubular steel and wood, intended for mass production, 1925. They were produced by the joinery workshop at the Bauhaus.

Some workshops – for example, the wallpaper department, the weaving and furniture workshops and the metal workshop – were extremely profitable, and the new advertising department was awarded a contract to design newspaper advertisements for the company IG Farben. Meyer was dismissed by the local council in 1930 because of his Marxist beliefs and the excessive politicization of the school. Mies van der Rohe succeeded him in 1930, re-establishing the school's reputation. He promoted the teaching of architecture, relegating to the background any ambitions for social and cultural experiments. The Bauhaus became a technology college providing professional qualifications. Mies van der Rohe designed and made the Barcelona collection of tubular steel furniture, presented at the German pavilion of the 1929 International Exhibition in Barcelona. In 1931, when the Nazis won a majority in the Dessau parliament, the school's grant was cut, and it closed in 1932. Mies van der Rohe then rented a factory in a suburb of Berlin, from where he tried to relaunch the Bauhaus. He turned it into a completely private, fee-paying institution, which he funded by selling the patent rights on the objects or furniture designed at the school. By this course of action he hoped to ensure the school's survival, but on 11 April 1933 the police turned up and closed the school down. In August, Mies van der Rohe announced that the Bauhaus had been dissolved.

Paris: an inventory

The International Exhibition of Modern Decorative and Industrial Arts – initially planned for 1915, but postponed until 1916 because of the war, and then again to 1922 and 1924 – was finally held in Paris in 1925. France and 21 other countries were represented. The aim of the exhibition was as follows: 'Through collaboration between artist, industrialist and craftsman, to bring together in an international exhibition all the decorative arts: architecture, woodwork, stonework, metalwork, ceramics, glass, paper, fabrics, and so on, in all their forms, whether applied to utilitarian objects or luxury items, and whatever their purpose: the exterior or interior decoration of public and private buildings, furnishing, personal adornment, and so forth. This exhibition will show only modern art, and no copy or pastiche of old styles will be accepted.'

This event, which was intended to encourage 'social art', turned into a celebration of the roaring twenties – a bright contrast to the dark years of World War I, when the extravagances of Art Nouveau had finally burnt themselves out. It reflected the image of a deeply scarred society trying to make the most of the present.

The International Exhibition of Modern Decorative and Industrial Arts

The exhibition included prestigious pavilions such as the 'Collector's House' by Jacques-Émile Ruhlmann, an interior decorator assisted by Francis Jourdain and Henri Rapin. It was furnished with objects designed by Brandt, Puiforcat, Lecoeur and Lenoble. The interior contained high quality cabinet-making that won favour with the public. The pavilion designed by the Compagnie des Arts Français, presided over by Louis Suë and André Mare, was described as 'a museum of contemporary art'. It exhibited furniture made of gilded wood, macassar ebony and rosewood constructed in the Louis-Philippe tradition; the pictures were painted by friends of the group: Marie Laurencin, Charles Dufresne and Dunoyer de Segonzac. The flagship pavilion of the Société des Artistes Décorateurs (SAD), called 'A French Embassy', was an imposing creation which hardly fulfilled the original brief given: 'to create inexpensive, truly democratic art'. Only the smoking-room by Jean Dunand was modern in spirit. The group's interior decorators worked for the elite, and they were benefiting from the development of private villas and town houses, for which they created masterpieces that were the pride of the French luxury market. The SAD was rapidly expanding. It left its cramped premises in the

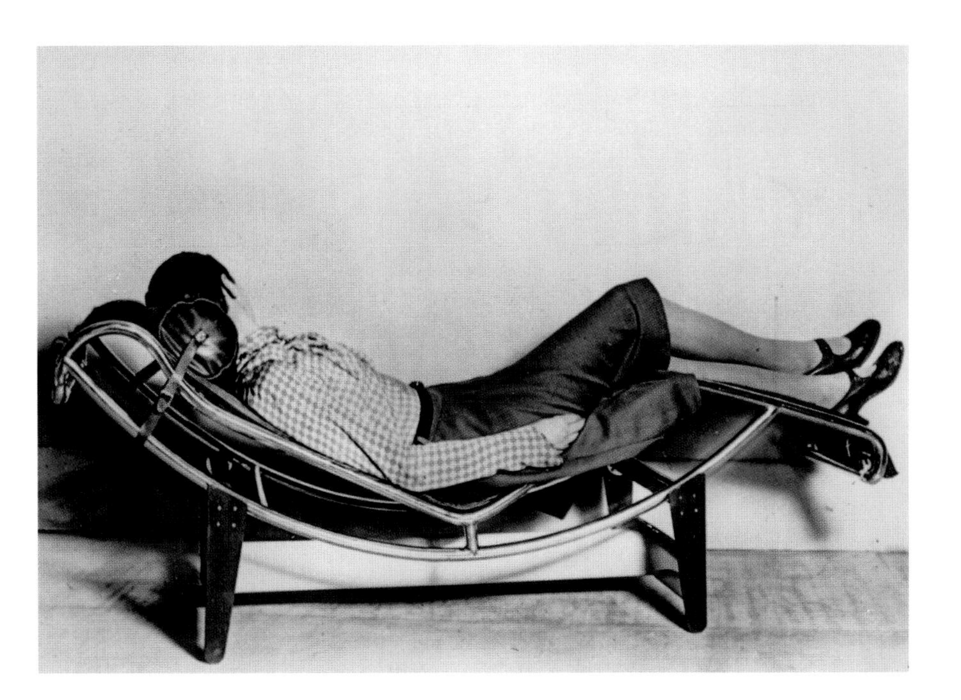

LC4 chaise longue with a metal frame in the form of an arc of a circle, chrome-plated tubular steel and leather, Le Corbusier, Pierre Jeanneret and Charlotte Perriand, 1928. Paris, Le Corbusier Foundation.

Pavillon de Marsan to take up quarters in the Grand Palais. Pierre Chareau belonged to the modern faction in the SAD.

Department stores: showcases for the exhibition

The four pavilions run by department stores were a central focus of the exhibition. They showed interiors which were extremely attractive but which did not come within the budget of the average consumer. The studios of the department stores approached their brief in various ways. The Studium Louvre presented a modernist image with designs by Djo-Bourgeois. At the opposite end of the scale, Paul Follot came up with a rich, ornamental, overloaded style for the Pomone studio of Bon Marché. The Primavera studio of the Le Printemps department store and the drawing office of Galeries Lafayette favoured a more restrained style.

These studios were trying to promote a successful collaboration between art, industry and business. They wanted furniture design to be more directly aimed at the general public, with the use of window displays, advertising and sales catalogues. However, their approach was still close to craftsmanship.

The 'New Spirit' pavilion

The word 'modern' was fashionable. In his article 'L'Exposition des Arts Décoratifs et Industriels de 1925, les Tendances Générales' in *L'Art Vivant* (1925), Waldemar Georges wrote: 'French and foreign interior decorators work only for a privileged class. None of the pavilions is aimed at the working-class public ... Modern architecture – the glass and steel architecture that held sway in 1899, and the reinforced concrete buildings of today – makes it possible to give an impression of strength, precision, lightness and transparency due to the economy of the materials used ... Apart from Perret's theatre, the two pavilions by Robert Mallet-Stevens and the USSR pavilion, the "New Spirit" villa is the only building in the whole exhibition that can be described as modern – that is, it fulfils its mission from both the practical and the aesthetic point of view.'

The 'New Spirit' pavilion was designed by Le Corbusier, Ozenfant and Pierre Jeanneret as an anti-decorative manifesto. Positioned well away from the centre of the exhibition, it was an affirmation of Le Corbusier's beliefs about architecture, town planning and interior design. The pavilion demonstrated the advantages of prefabrication. The main structure, on two levels, included a model flat, for which furniture was chosen according to functional rather than aesthetic criteria.

Further examples of modernity

Other outstandingly modern contributions were the pavilions of Lyons and Saint-Étienne, designed by Tony Garnier, and the cinema pavilion by Francis Jourdain (1876–1958). Jourdain had been interested in furniture since 1904 and had produced a series of interchangeable pieces. In 1912, he set up the Ateliers Modernes, manufactured his 'interchangeable' furniture and founded the group Art Urbain; Mallet-Stevens, Chareau and Gabriel Guevrekian became members. At the 1925 exhibition, Jourdain designed a smoking-room and gymnastics hall in the 'French Embassy' pavilion. Robert Mallet-Stevens (1886–1945) was greatly influenced by the Palais Stoclet, which Josef Hoffmann built for his uncle. Commissioned to build a villa for Viscount Charles de Noailles in Hyères, he approached Chareau, Guevrekian, Van Doesburg and Jacques Lipchitz; Van Doesburg designed the Flower Room. Mallet-Stevens was open to avant-garde influences and close to the Bauhaus and De Stijl movements. At the Salon d'Automne in 1924, he and his French collaborators – Jourdain, Jan and Joël Martel, Blanche Klotz and Jean Perzel – exhibited an interior. Then, for the 1925 exhibition, he designed the entrance hall of 'A French Embassy', the tourism pavilion, the pavilion of the Paris tourist office and the garden with concrete trees by Jan and Joël Martel.

Some foreign pavilions were strikingly modern, including the Belgian pavilion designed by Victor Horta, the Czechoslovakian pavilion by Josef Gocar (the master of Czech Rondo-Cubism), the Austrian pavilion by Josef Hoffmann, the Italian pavilion with its Futurist offerings, and Konstantin Melnikov's Soviet pavilion – far and away the most innovative, with schemes by Vesnin, Tatlin and Popova, and Rodchenko's Workers' Club, which could be dismantled and taken elsewhere. The Soviet pavilion was awarded first prize. Germany and the USA were not represented at the exhibition.

Le Corbusier

Exhibition stand at the Salon d'Automne, designed by Le Corbusier, 1929. *LC4* chaise longue, *Grand Confort* easy chairs, swivelling armchair, tables with metal stands and glass tops, modular filing system. Paris, Le Corbusier Foundation.

The Swiss architect Charles Édouard Jeanneret-Gris, better known as Le Corbusier (1887–1965), completed his training in the Florence region of Italy and in Paris, where he worked in the practice of the Perret brothers (1908–9). He designed his first pieces of furniture for his parents' house at La Chaux-de-Fonds in 1912. In 1915 he invented the 'Domino House', based on a standard module. In early 1917, at the age of 30, he took up residence in Paris.

In 1919, together with Amédee Ozenfant and Paul Dermée, he founded the magazine *L'Esprit Nouveau*, which promoted his ideas and provided a platform for other movements. In 1921, he and Ozenfant published the Purist manifesto under the title *Après le Cubisme*. Le Corbusier opened his own practice on the Rue de Sèvres in Paris with his cousin Pierre Jeanneret in 1922. In *L'Art Décoratif Aujourd'hui* (1925), he came up with the terms *machine à habiter* (machine for living), *équipement de la maison* (household fittings) and *machine de repos* (machine for rest), intended to replace 'house', 'decoration' and 'seat' respectively. His standard modules – elements made from wooden half-cubes that could be juxtaposed and superimposed – made their first appearance in the 'New Spirit' pavilion at the 1925 International Exhibition of Modern Decorative and Industrial Arts.

Le Corbusier devoted much thought to town planning and admired the structural beauty of cars and liners. At the 1929 Salon d'Automne, he exhibited the fully furnished interior of a house. Almost all of the furniture was made in collaboration with Charlotte Perriand. The *Grand Confort* easy chair (large and small model) was designed in 1928, and the self-adjusting chaise longue with a supporting base was produced the same year; another chaise longue model, the swivelling armchair and the aeronautical steel tube table date from 1929. Le Corbusier established a system of proportions, known as the Modulor, which was based on the golden mean and which he applied to his architecture and furniture.

The Union des Artistes Modernes (UAM)

The Union des Artistes Modernes officially came into being on 15 May 1929 after a small breakaway group reacted against the conservatism of the Société des Artistes Décorateurs. This group, consisting of René Herbst, Djo-Bourgeois, Jean Luce, Jean Puiforcat, Georges Fouquet and Charlotte Perriand, wanted to display jewellery, ceramics, advertisements, everyday objects, and so on, alongside furniture and traditional decorative objects. They exhibited at the Salon d'Automne in 1929. According to the first article of their articles of association, the objectives of the group were 'to bring together like-minded artists with the same spirit and the same attitude to trends, combine their efforts, and ensure they reach the public by means of an annual international exhibition in Paris and a newsletter'.

With Mallet-Stevens as president, the managing committee consisted of Herbst, Jourdain, Hélène Henry, and Raymond Templier. The founding members were French, or foreigners living in France, such as the Hungarian sculptors Joseph Csaky and Gustave Miklos or the Armenian architect Guevrekian. They held their first exhibition in 1930 at the Musée des Arts Décoratifs. They accepted items by foreign artists, such as Rietveld and Van der Leck of the De Stijl movement. The second exhibition in 1931 was open to German, American and British artists. Gropius, the former director of the Bauhaus, took part as an active member of the UAM. There were many affinities between the UAM and the Bauhaus, both of which were made up of professionals in the fields of architecture, graphic art and decoration, as well as craftsmen, engineers and artists. The UAM was helping to bring modern designers together, and when the Viennese architect Josef Margold and the German architect Alfred Gellhorn expressed their wish to belong to the group, they were invited to join. The centrepiece of the 1932 exhibition was the display of work by Soviet and Czech graphic designers organized by Jean Carlu.

Refectory chairs, Robert Mallet-Stevens, 1926. Paris, Georges Pompidou Centre.

Cantilevered chair made from gas pipes, Mart Stam, 1926. Vitra Design Museum.

The same year saw the arrival of a new member, the Italian Alberto Sartoris, while in 1933 it was the turn of the Dutchman Bernard Bijvoet. In 1934, they drew up their first manifesto, *Pour l'Art Moderne, Cadre de la Vie Contemporaine*, with contributions by the critic Louis Chéronnet and by Jean Carlu, who detailed their objectives and future activities. They advocated mass production, the exploitation of technological progress, and the use of certain materials: 'the quartet of concrete, glass, metal and electricity – a harmonious quartet, whose principles and relations one to another we have tried to establish'. These architects made use of all the arts and crafts without exception: stained glass, graphic art, advertising and lighting, as well as decorative sculpture and painting. It was the UAM's goal to forge close links with industry, and in 1934 this led to the setting up of a permanent committee overseeing contacts between the UAM and the OTUA (Office Technique pour l'Utilisation de l'Acier). Designers of forms and producers of materials, the members joined forces under the direction of Herbst to come up with schemes for fitting out liners. The all-important idea of modernity was expressed in particular by the search for new materials. Jean Burkhalter had the idea of using a string mesh for the seat of his tubular chair, while Herbst used the elastic cord normally used in physical fitness equipment and stretched it across the seat of his chair.

In his masterpiece the *Glass House* (1931), Pierre Chareau (1883–1950), assisted by Bijvoet and the ironworker Louis Dalbet, showed his liking for using glass as a façade, as well as for metal locks and joinery elements, for mobility, for the evolutionary character of

The Noailles villa at Hyères (Var), Robert Mallet-Stevens, 1924–33. Built in stages for Charles and Marie-Laure de Noailles, the house is, in a cubist spirit, a 'mass of grey cement squares', with pool, terraces and hanging gardens.

Boeing

The 'Monomail', 1929.
Ce monoplan tout en métal est l'un des premiers avions de ligne mis au point par Boeing,

William E Boeing (1881–1956) was educated at Yale University, then worked in the timber industry – an experience that proved useful to him when he came to design planes. An aviation enthusiast, he trained as a pilot before going into partnership with George Conrad Westervelt, an American naval engineer with aeronautical training, to produce the *B&W* – a plane constructed from wood, canvas and iron – in 1916. Westervelt rejoined the army, and the Boeing Airplane Company then took on Tsu Wong, an aeronautical engineer, to design the *Model C*, which was tested and purchased by the American military. In 1923, Boeing's *Model 15 (PW-9)* was again chosen by the army, being developed over the following decade to become the *Model P-12/F4B*. Charles Lindbergh's outstanding feat of flying non-stop from New York to Paris in 1927 fired the public's imagination. The popularity of passenger flights inspired Boeing to come up with the *Model 80* three-engine biplane for 12 passengers, which made its first flight on 27 July 1929, and then the *Model 80A* for 18 people. One of the first airliners, perfected by Boeing that same year, was the 'Monomail', an all-metal monoplane intended for the transport of mail and freight. This was the basis for the Boeing *247*, which United Airlines put into service in 1933. For Pan American World Airways, Boeing developed a commercial aircraft to transport passengers across the Atlantic. The first flight of the Boeing *Clipper 314* took place in June 1938. It was the largest civilian aircraft of its time, capable of carrying 90 people on daytime flights and 40 people on overnight flights. A year later, the first regular service between the USA and Britain was launched.

spaces, for fold-down, adjustable, transformable forms, and for small wrought-iron tables arranged in a fan shape. The *Glass House* encapsulates his beliefs as an architect and interior designer.

Jean Prouvé (1901–84), himself a designer, put all his manufacturing talents at the service of some of the notable figures in the UAM, including Perriand, Georges-Henri Pingusson and Marcel Gascoin. In 1937, the UAM was asked to devise an exhibition programme with a scientific and technological slant. The group backed the pavilion by Le Corbusier called 'Modern Times', securing a site on which to build it. Constructed in ten weeks by Pingusson, Jourdain and André Louis, it included prefabricated elements. An exhibition of commonplace objects and furniture was presented, thus satisfying Jourdain's desire for a 'bazaar' of everyday life. Glass was featured as a 'modern' material used in a rational context. The window openings of the UAM pavilion were so large that they diverted attention away from the supporting structures. René Coulon, who was also involved in the 'Hygiene' pavilion and the Saint-Gobain pavilion, did not join the UAM until 1944. A consultant architect to the Saint-Gobain glassworks, he was the first person to perfect 'transparent concrete' – glass tiles mounted in concrete. He also produced interesting pieces of furniture from moulded toughened glass. Louis Sognot created basketwork furniture.

Herbst succeeded Mallet-Stevens as president on the latter's death in 1945. In 1949, he staged an exhibition entitled 'Formes Utiles, Objets de Notre Temps', which brought into being the Formes Utiles section, an offshoot of the UAM. This marked the start of disagreements that led to the break-up of the group in 1958.

Industrial design in the USA

The first exhibition of modern decorative arts, the 'Werkbund Exhibit of Industrial and Applied Arts', was held at the Newark Museum in New Jersey in 1912 at the instigation of John Cotton Danna. Fascinated by German design, which he knew through the press, he mounted an exhibition of 300 items, including photographs of buildings designed by Behrens and Gropius.

The impact of 1925

Like all major countries, the USA was invited to participate in the International Exhibition of Modern Decorative and Industrial Arts in Paris in 1925. However, it had to demonstrate that it could manufacture innovative products that would sell. The key word was 'modern'. Herbert Hoover, the American Secretary of State for Trade, held a consultation involving teachers, businessmen, craftsmen, manufacturers and leading figures in the American art world and reached the conclusion that there was no modern design in the USA. Feeling it was best to decline the invitation, he sent Charles R Richards to Paris. In his report on his visit, Richards took the view that the Paris exhibition had not really been inspired by the spirit of modernity, and that most of the French exhibits were in fact reinterpretations of styles derived from the past. Nonetheless, the 1925 exhibition was a great success, attracting many American tourists, including some future leaders of American design.

Showing the way

The Americans did not seem ready to cross the threshold into modernity. However, Joseph Breck, Richard F Bach and Charles C Richards organized annual exhibitions of objects produced by American factories at the Metropolitan Museum in New York, as well as showing European designs in the hope of encouraging innovation. In 1926, the museum hosted two major exhibitions: the 'Exhibition of Current Manufactures Designed and Made in the USA' and, a few months later, a travelling exhibition of a selection of objects from the International Exhibition of Modern Decorative and Industrial Arts. Continuing with this dynamic policy, the museum organized an exhibition of contemporary Swedish decorative art in 1927.

Yet another exhibition was held in 1927: the 'Machine Age Exposition', organized by Jane Heap in office space on 57th Street in New York over a two-week period. The committee included Alexander Archipenko, Charles Demuth, Marcel Duchamp, Hugh Ferris, Louis Lozowick, André Lurçat, Elie Nadelman, May Ray and Charles Sheeler. The catalogue cover was by Fernand Léger. Products were exhibited that celebrated the 'dynamic beauty' of machines; tribute was paid to the engineer rather than the decorator. The items shown – drawn from the USA, Austria, Belgium, France, Poland and Russia – anticipated the exhibition 'Machine Art', held at the Museum of Modern Art in New York in 1934.

In 1928, the American Designers Gallery brought together the work of 36 exhibitors, including Paul T Frankl, Raymond Hood, William T Lescaze and Joseph Urban. For the first time a group of American designers were exhibiting works that expressed an American identity. Ten complete interiors were displayed, including Donald Deskey's smoking-room, which was made from industrial materials, with an aluminium ceiling, dark linoleum on the floor and aluminium lights. The use of materials such as cork, aluminium and metal was characteristic of the modern American style. In June 1928, the magazine *Good Furniture* joined forces with artists and published articles about modern products. Frankl gave expression to modernity by producing furniture with 'skyscraper' aesthetics. The decorative arts, painting and photography also drew on the aesthetics of verticality for inspiration. After the eccentricities of the 'skyscraper' style, a more restrained 'streamline' trend took over. Lines were inspired by speed; but designers were not content to

The Streamline Style

Airline armchair, maple and leather, Kem Weber, 1930. An armchair with aerodynamic lines. Houston, Museum of Fine Arts.

Speed and the dynamic beauty of machines caught and fired designers' imaginations, as the Italian Futurists had predicted in 1909. The Streamline Style appeared around 1930. Norman Bel Geddes championed it in his book *Horizons* (1932). Streamline combined the principles of aerodynamics with the functional geometry of the International Style. Although the style first surfaced in Germany, in the form of Count Zeppelin's airship, it was American designers who developed its aesthetic aspect. Aerodynamic research was carried out using wind-tunnel trials; the results helped to produce more compact, curved, missile-like monocoque forms, especially in the field of transport (planes, ships, cars and trains). New techniques helped to bring these changes about: for example, the forming machine for new materials such as stainless steel, polished aluminium sheets, Bakelite and plastics, which made it possible to produce moulded objects in innovative shapes that reflected a vision of a future society.

The Streamline Style with its glamorous overtones became popular in the USA in the 1930s, giving a boost to consumerism. Its success can be explained not only by people's desire for speed and efficiency but also by subliminal sexual references. A debate took place between the Streamliners and the machine purists, who were championed by the Museum of Modern Art. Streamlined forms made their appearance in furniture and clocks designed by Gilbert Rohde (1894–1944) and produced by Herman Miller; Kem (Karl Emanuel Martin) Weber (1889–1960) produced vases, cocktail shakers and furniture; and Walter Dorwin Teague (1883–1960) designed cameras for Kodak, then worked for the Sparton Radio Company. Warren McArthur was the first person to make anodized aluminium furniture that combined artistic craftsmanship and machine aesthetics.

apply these principles only to railway engines and transatlantic liners – they extended them to everyday objects.

Department stores: the Parisian model

Department stores had played an important role in the Paris exhibition of 1925; now the same thing happened in New York. Macy's store organized an 'Exposition of Art in Trade' in spring 1927, presenting American and French designers and showing an entire room of furniture in Frankl's 'skyscraper' style. In 1928, the department store Lord & Taylor held 'An Exposition of French Decorative Art', including fine objects and pieces of furniture by Ruhlmann, Dunand, Chareau, DIM (Décoration Intérieure Moderne) and Rodier. It was a resounding commercial success. In 1928, Macy's organized a second exhibition, 'International Exposition of Art in Industry', presenting a wide range of design from Austria, France, Germany, Italy, Sweden and the USA. However, in 1929 came the Wall Street crash, marking the beginning of an economic depression with unprecedented worldwide repercussions.

Maquette of the city of *Futurama*, designed for General Motors by Norman Bel Geddes, 1939. Displayed at the 'World of Tomorrow' exhibition in New York.

The International Style

In 1932, Philip Johnson and Henry-Russell Hitchcock organized an exhibition entitled 'Modern Architecture: International Exhibition' at the Museum of Modern Art in New York. For the first time they identified a shared vocabulary, an 'international style' applied to architecture and modern design; its representatives included Mies van der Rohe, Oud, Le Corbusier and Frank Lloyd Wright. The Department of Architecture and Industrial Art was established at the Museum of Modern Art that same year. The USA had caught up with the European avant-garde. The 'Century of Progress' exhibition in Chicago in 1933–4 – dominated by the Goodyear airship and the superb Zephyr train, made by the Burlington & Quincy Railroad company, and its rival the M-10000, made by the Union Pacific Railroad – presented a positive image of science and progress. This recognition of the value of science and trade was supported by investment from the National Research Council (NRC). Advertising helped to popularize scientific idealism and a vision of a future shaped by science. Sales catalogues, advertisements in magazines and product packaging were designed by distinguished names such as A M Cassandre, José Arentz or Otis Shepard.

The USA wakes up

The USA was roused to action by the New Deal implemented by Franklin D Roosevelt in 1933, which looked to designers to help relaunch the economy. Some new players appeared on the scene: industrial designers, who had backgrounds in the visual arts, fashion and publicity.

The most flamboyant of these was Norman Bel Geddes (1893–1958), an advertising agent, stage designer and architect, who had also

Raymond Loewy

Raymond Loewy perched on a K4 railway engine, redesigned for the Pennsylvania Railroad Company, 1936. A superb example of aerodynamism.

Raymond Loewy (1893–1986) was of French origin, but lived in the USA. In 1919, he was working as a fashion illustrator, and had to eke out a living during the years of economic recession. His message to industrialists was that aesthetics was a selling point. He was finally listened to, and in 1932 he opened his first agency in New York. He was given design commissions by Sears Roebuck and Company, the Greyhound Corporation and the Pennsylvania Railroad Company. The *Designers Office and Studio* interior by Loewy and Lee Simonson featured in the 'Contemporary American Art' exhibition at the Metropolitan Museum of Art. In 1934, Loewy opened an office in London to handle a commission for duplicating machines from Sigmund Gestetner. In 1936, the naval architect George C Sharp joined forces with him to design a modern ship for the Panama Railway Steam Ship Company. In 1937, he incorporated a department of architecture and interior design into his New York agency. At the 1937 International Exhibition of Arts and Technology in Modern Life held in Paris, he was awarded the gold medal for the GG1 railway engine.

The opening of his Chicago agency in 1938 marked the beginning of his collaboration with Studebaker and Coca-Cola. He took part in the New York exhibition of 1939 and produced the exhibition buses for the Greyhound Corporation; he was also the consultant for the railroad pavilion, the largest in the show. In 1941, Loewy designed the new packet for Lucky Strike cigarettes, and devised a 'corporate identity program' for the International Harvester Company. In 1946, he became president of the Society of Industrial Designers, and in 1952 he founded the CEI (Compagnie d'Esthétique Industrielle) in Paris. He published *Never Leave Well Enough Alone* in 1951, arguing that 'ugliness doesn't sell'. He was keen to reconcile progress with human nature, and thereby encourage people to acquire goods for pleasure rather than out of necessity.

designed window displays on New York's Fifth Avenue. He described himself, in 1927, as an industrial designer whose projects were inspired by industry. In 1928, through his wife, he met J Walter Thompson, the chairman of the famous advertising agency, who commissioned him to design a conference hall that would function as efficiently as a machine. Thompson also introduced him to clients such as Simmons (a bed manufacturer) and Toledo (a maker of weighing machines). More than anyone else, Geddes was responsible for spreading the idea of streamlining. His work, which mainly consists of projects rather than implemented schemes, had a decisive influence. In 1932, he published the book *Horizons,* a manifesto for the Streamline Style. 'Speed is the cry of our era, and greater speed one of the goals of tomorrow.' He demonstrated that modernism and social art were compatible with the approach and objectives of industrial aesthetics.

Designers and products

Walter Dorwin Teague (1883–1960), an advertising agent, opened an agency in New York. In 1911, he travelled to Europe, where he discovered Le Corbusier, Mallet-Stevens and Gropius. On his return he designed the Baby Brownie camera (1927) for Kodak, which was a huge success. Collaboration with Kodak followed. He designed cars for Marmon in 1932–3, the Ford Motor Company's pavilion at the Century of Progress exhibition in Chicago in 1933, and a Ford pavilion for the San Diego Fair in 1935. In 1936, he was appointed a member of the organizing committee for the New York 'World of Tomorrow' exhibition, held in 1939. In 1940, he published *Design This Day – the Technique of Order in the*

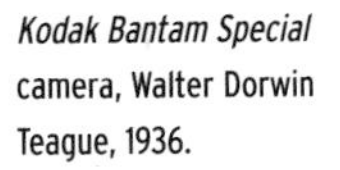

Kodak Bantam Special camera, Walter Dorwin Teague, 1936.

Machine Age, in which he argued that geometry is the essence of design, quoting the Parthenon in Athens as an example. He was an excellent businessman, giving the profession of industrial designer the serious image it needed if it was to win over sceptical industrialists. He was willing to be guided by fashion and to go along with the stylistic renewal demanded by his clients, convinced that what was good for business was good for society as a whole.

The designer Henry Dreyfuss (1904–72) was instrumental in founding the Industrial Society of America, and worked on industrial design for Hoover, John Deer, Hyster and Pan Am. He also designed the interiors of liners and planes. He worked for the Bell Company for 40 years, perfecting the famous *Model 300* telephone (1937), followed by the *Model 500* (1949). He refused commissions in which the engineer was not involved at the design stage.

The Cranbrook Academy

Eliel Saarinen (1873–1950), a Finnish architect, arrived in the USA after winning second prize in the competition for the Chicago Tribune Tower in 1922. He was already 50, and had behind him a long career in Finland and Europe, where he had produced work in the spirit of the Arts and Crafts Movement. In 1929, he showed a dining-room in the exhibition 'The Architect and the Industrial Arts' at the Metropolitan Museum in New York. When the businessman George G Booth decided to set up an art school on his land at Bloomfield Hills in Michigan in 1927, he met Saarinen (on Albert Kahn's advice) at the University of Michigan and asked him to come up with some ideas for the project, with regard to both architecture and course content. It took several years to construct all of the buildings. The students were involved in the gradual development of the school, in the fields of both architecture and design. A workshop of decorative sculpture run by Geza Maroti and, later, workshops for furniture, bookbinding and weaving were set up to complete the apprentices' training.

In 1928, Saarinen started on the construction of studio premises and accommodation for the students and teachers. Booth himself was overwhelmed by the scope of the project and so appointed Saarinen president of the Cranbrook Academy of Art in 1932. In spite of the Academy's economic difficulties, the spirit that prevailed there was unique: the individual development of students was encouraged in a wide variety of skills, and there was no pre-established curriculum. The community was enriched by collective experiments, festivals, exhibitions and lectures by famous architects such as Le Corbusier and Alvar Aalto. Saarinen's mission was accomplished. Within just a few years Cranbrook became a prestigious training centre for students destined to make their name in the arts, such as Charles Eames and Florence Knoll, who both studied there in the 1930s. This unique experiment represented an important chapter in the history of American design.

Teachers at the New Bauhaus (which became the School of Design in 1939). The New Bauhaus was a school founded in Chicago by the Hungarian painter and sculptor László Moholy-Nagy in 1937.

Ferment in Europe

The increasingly unstable political situation in Europe brought waves of immigration to the USA, resulting in a virtual globalization of the Modern Movement. After the closure of the Bauhaus in 1933, a large number of architects, designers and artists – some of German origin – left Europe. Nikolaus Pevsner published *Pioneers of the Modern Movement from William Morris to Walter Gropius* (1936), which described the development of the Modern Movement.

The majority of European expatriates arrived on the other side of the Atlantic from 1937 onwards. First of all came those from the Bauhaus: Gropius, Breuer, Moholy-Nagy and Hin Bredendieck. Mies van der Rohe took up residence in Chicago, where he ran the school of architecture at the Illinois Institute of Technology (IIT). Siegfried Giedion became a lecturer at Harvard, embarking on a distinguished career. The development of modern architecture and design in the USA was turned upside down by the teaching that they did within influential institutions and by the practical projects they undertook. The Museum of Modern Art exhibition devoted to the Bauhaus (1938) was staged by Bayer, Gropius and Albers. It presented the philosophy and work of the school to the

American public, prompting an instant infatuation with metal furniture: the *Beta Chair* armchairs (1930) by Nathan George Horwitt, aluminium armchairs by McArthur, and the *Streamline* armchair by Weber.

The New Bauhaus

The New Bauhaus was founded in 1937 by the Chicago Association of Arts and Industries, which invited Moholy-Nagy to run it. The curriculum was very much inspired by the Bauhaus, but more emphasis was placed on professionalism than on technical and formal experiment. Moholy-Nagy was helped by one of his former pupils, Hin Bredendieck, by the Russian sculptor Alexander Archipenko, who had emigrated to America in 1923, and then by Jean Hélion and Herbert Bayer. Conflict with the school's sponsors led Moholy-Nagy to found the School of Design in 1939. He added courses supporting the war effort to its programme: a course in camouflage and a course for the war disabled; these attracted sizeable subsidies. The School of Design became the Institute of Design in 1944.

The birth of marketing

Consumerism became a real social phenomenon in the 1920s. People began to think up new advertising and marketing ideas. The development of mass production and the rational organization of business had brought about a change in commercial practices. In 1908, the Ford Motor Company launched the *Model T*, which was a huge commercial success. They set a trend by developing a new model every year. Ford was regarded as the inventor of a new theory of industrial organization and activity, seen by some economists as an economic system in its own right. But this new business 'religion' in the USA provoked criticism, especially from intellectuals such as Lewis Mumford, Stuart Chase and John Roderigo Dos Passos; some left for Europe. In 1927, Macy's department store in New York organized an exhibition called the 'Exposition of Art in Trade'.

The year 1930 marked the publication of the first issue of *Fortune* magazine. The man behind the magazine was Henry Luce, who produced it for the heads of industry and not for the workers. He employed the best writers and photographers. Industrial companies were analysed, and new phenomena such as modern architecture and industrial design were examined. It looked as though the USA would be able to bring together designers and industrialists, something neither the Werkbund nor the Bauhaus had really succeeded in achieving. The aesthetic factor became an extra selling point. American industries secured the help of management teams and marketing teams, which supplied designers with the information they needed for their work: market studies, investment data, and so on. (Sometimes, however, they were excessively influenced by this and failed to think sufficiently about form – 'styling' could lead them astray aesthetically.) Products sold not only for their technical qualities but also for their appearance and style. The idea of 'progress' became a factor in product design, especially during the recession of the 1930s, when business turned to science and research. New 'heroes' such as Walter P Chrysler and David Sarnoff appeared on the scene.

Scandinavian design, organic design

Until the late 19th century, the Scandinavian countries depended mainly on agriculture. Then came the period of National Romanticism, when a great many historical museums were opened, such as the Atheneum in Helsinki. This museum was linked to a school of decorative arts which was the cradle of Finnish design. The Finnish Association of Arts and Crafts and the Swedish Handicraft Association were also active. Finland made its mark at the Paris World Fair in 1900 with its national pavilion, designed as a total work of art by the architect Eliel Saarinen. With this contribution, Saarinen brought Scandinavia back into the mainstream of European culture.

Armchair 41 (or Paimio Chair), Alvar Aalto, 1931–2. Laminated beech frame and preformed plywood seat.

Alvar Aalto

Stacking stools with L-shaped legs, Alvar Aalto, 1931–2.

Alvar Aalto (1898–1976) was a Finnish architect who designed buildings in the Neoclassical style. In 1924, he married Aino Marsio, also an architect. He made his first items of furniture for his brother in 1921, then he and Aino were commissioned to fit out the interior of a large café at Jyväskylä. In the summer of 1924, they fitted out the premises of the Häme association, before winning a competition organized by the Finnish Society of Decorative Arts: the remit was to furnish the living-room of a family on a modest income.

Alvar Aalto travelled widely, meeting the architect Erik Gunnar Asplund in Stockholm, and then Sven Markelius. Aalto and his wife made their home in Turku (Finland) in 1927. He abandoned Neoclassicism in favour of the International Style and designed the head office of the daily newspaper *Turun Sanomat* (1928–9) and the Paimio sanatorium (1929–33). Together with Otto Korhonen, the director of the Huonekalu factory, he tackled the problem of how to make mass-produced furniture. His first chair was a stacking one, combining solid wood and preformed plywood, and was assembled by hand. For mass production he came up with a prototype with a tubular steel frame. Various technical experiments led him to discover new forms for wood. Aalto made legs of laminated beech, screwed on under an incurving seat; the legs then extended up to become arm-rests. In 1932, he perfected the famous Aalto underframe and designed many types of seating for the Viipuri library, which was officially opened in 1935. The Artek company, founded in 1935, sold and distributed his models internationally. The furniture was made from natural materials; the forms were ergonomic, but still austere and simple.

An exhibition of Aalto's work was held in New York in 1938, and in 1940 he went back to the USA, where he gave lectures on reconstruction in Finland. He started to teach at the Massachusetts Institute of Technology the same year.

A democratic way of living

In Scandinavia more than anywhere else, designers took a democratic approach to design with the objective of producing an ideal society and improving the quality of life. Scandinavian design was sustained by a humanist ethic whose origins can be traced back to Lutheranism, the state religion in the five Scandinavian countries of Denmark, Finland, Iceland, Norway and Sweden. For centuries the family home had been the focal point of life for Scandinavians – a vital safe haven against often hostile climatic conditions and the framework for family life. People had learnt to manage the available resources and use them with maximum efficiency. Design was part and parcel of their cultural, economic and social life. The combination of ancestral skills and modern techniques meant that Scandinavians were able to produce high-quality objects that balanced form, function, colour, durability and cost.

Until the 1950s, Denmark had very little industry; its economy depended on agriculture and traditional crafts. The anthropometric research carried out by Kaare Klint (1888–1954) in 1916 had a huge influence on design. As a result of detailed studies of anatomical proportions, he produced furniture such as the *Safari* chair and the *Deck* chair in 1933. However, he was not interested in the industrial exploitation of his designs. The Fritz Hansen company, which specialized in turned wood, devoted itself to producing designers' furniture, thus helping to spread Danish design on a large scale. Poul Henningsen (1894–1967) was interested in the technical aspects and aesthetics of lighting.

Finland set the trend for an interdisciplinary approach to architecture and design ahead of the Werkbund and the Bauhaus. The development of Finnish design was often linked to international exhibitions, as evidenced by Eliel Saarinen's pavilion at the Paris World Fair in 1900. Alvar Aalto pursued the quest for complementarity between form and function.

The construction of Stockholm Town Hall in 1923 showed what Swedish design could do. Many links were forged between Swedish artists and Swedish industry, especially in the areas of glass and ceramics, under the aegis of the society aimed at fostering Swedish handicrafts (founded in 1845). Axel Larsson exhibited mass-produced furniture with organic forms in the Swedish pavilion at the New York exhibition in 1939. The Svenskt Tenn company made furniture designed by the Austrian émigré Josef Frank (1885–1967). During the 1920s, many Swedish designers went off to study in Paris and Berlin. In the following years, luxury was replaced by Swedish functionalism; modern design was regarded as a tool for social change. Design was supposed to act as a social binding agent and turn the vision of the 'Folkhemmet' (Everyman's House) into a reality. Erik Gunnar Asplund came up with a

Interlocking aluminium tables, Friedrich Kiesler, 1935.

Table from the *Arabesque* series, Carlo Mollino, c.1950. Paris, Georges Pompidou Centre.

Low table, Isamu Noguchi, 1944. Glass table top and wooden underframe. Made by Herman Miller. Vitra Museum Design.

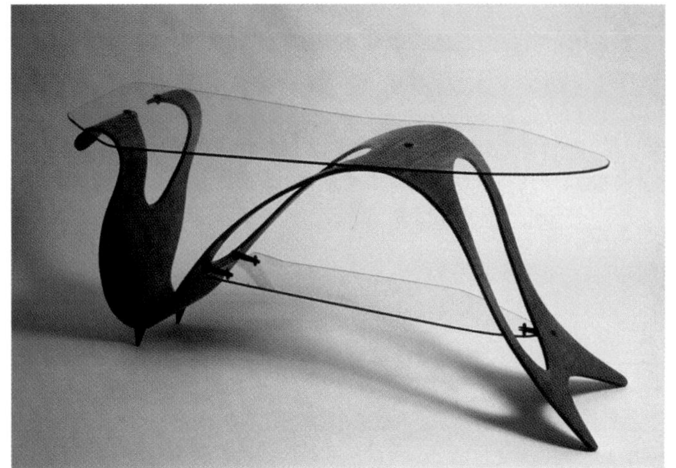

modern form that matched the functionalists' credo, according to which 'well-thought-out design is beautiful'. Bruno Mathsson designed chairs with organic forms using laminated wood and webbing.

Scandinavian architects travelled frequently between Scandinavia and the Baltic countries, but also continued the tradition of making study tours to central and southern Europe. In the early 1920s, their preferred destination was Italy, which was visited by Asplund and Sven Markelius in 1920 and Aalto in 1924. With the advent of functionalism in 1925–6, Scandinavian designers went instead to the countries where it had originated: Germany, the Netherlands and France.

Biomorphism

The origins of this movement can be found in painting and sculpture. It dates from 1915 and 1917, with Jean Arp and the Dada movement in Zurich. Then, in the 1930s, artists such as Salvador Dali, René Magritte, Joan Miró, Fernand Léger and Alexander Calder used similar forms. These ideas were also taken up by the British sculptors Henry Moore and Barbara Hepworth. Biomorphism emerged fully in the mid-1930s, when the Surrealist movement from which it derived was at its height.

The third dimension

Biomorphically inspired furniture is characterized by undulating forms echoing those of the human body.

It is exemplified by models such as Aalto's *Paimio* armchair (1931–2) or Mathsson's armchair (1933–6). Frederick Kiesler designed low aluminium tables shaped like amoebas (1935); Marcel Breuer designed plywood chairs with undulating forms such as the *Isokon Long Chair* (1935–6); and Isamu Noguchi made a table for the Goodyear Company that was in effect a functional sculpture. These examples show how design was developing from the two-dimensional to the three-dimensional. The style continued, becoming more marked after World War II, especially in the USA. In 1940, Bloomingdale's department store in New York, in association with the Museum of Modern Art, organized a pan-American competition on the theme of 'Organic Design in Home Furnishings'. Charles Eames and Eero Saarinen won first prize with an armchair inspired by the forms of the human body. For this they used new techniques and materials, including a preformed plywood shell. The Italian Carlo Mollino pushed biomorphism as far as it would go, drawing his inspiration from Surrealism and designing furniture with extraordinarily free forms. This trend spread to the decorative arts, particularly gold work, but also to architecture; this can be seen in 1950s buildings such as Eero Saarinen's TWA terminal at Kennedy Airport.

General view of the International Exhibition of Arts and Technology in Modern Life, Paris, 1937. The USSR's pavilion is opposite that of Germany.

Power and design

In the years leading up to World War II, there was a desire for a return to order in the air. There were more and more examples of historical reference – colonnades, pediments and so on – and at the same time there was a return to regionalism, noticeable especially in architecture. The Modern Movement had become suspect. But experiments continued, and industrial design was used as a propaganda tool, as exemplified by Volkswagen's Beetle car in Germany.

Totalitarian regimes

In Italy, Mussolini was not particularly interested in the arts, but he encouraged coordination between the state, artists, craftsmen and intellectuals. The theme of the Monza Triennale in 1930 was electricity in the home. Following in the footsteps of the Futurists, Mussolini wanted to make his Fascist regime synonymous with speed and machines, and looked down on any country that was stuck in the past. He adopted the slogan of the Futurist poet Emilio Marinetti: 'Velocizzare l'Italia' (Speed Italy Up). Modern industrial design was adopted as an official style, in preference to handicrafts.

In the USSR, Lenin put forward a utopian vision of society: 'Communism is Soviet power plus the electrification of the whole economy.' The power station, with its huge dimensions and capabilities, became the emblem of the new state. Stalin succeeded Lenin in 1924. After his dictatorship had become established and following the great purges of 1935, he imposed his artistic views on the country, choosing the tractor as the linchpin of his programme to promote collective agriculture.

His five-year plan also proposed a big increase in steel production, and his policy of building large residential blocks, hotels and ministries produced a style that combined classicism and modernity.

Hitler came to power in Germany in 1933. His totalitarian regime was based on visual display, preferring the Neoclassical style in both architecture and decoration. Hitler chose Gothic lettering as the official typeface, thus asserting the German origin of typography as represented by Gutenberg's printing press. Interior furnishings paid tribute to the qualities of German craftsmanship, in which traditional raw materials – wrought iron, wood and fabric – were used. He rallied the artistic establishments of Hamburg, Munich and Dresden, and opposed 'Jewish art'. Some schools were closed, the Bauhaus among them. Newspapers were banned. Some artists were deported, while others left the country of their own accord. Himself a disappointed artist, Hitler made up for it by staging huge propaganda spectacles inspired by Wagner's operas. He invented 'son et lumière' for his grandiose night-time spectacles. Music, sculpture and architecture fired him with enthusiasm, and he exploited them in his capacity as Führer.

The 1937 exhibition in Paris: between classicism and modernity

The building of the Palais de Chaillot in Paris for the 1937 World Fair revealed a taste for monumental, sober classicism – the officially approved style. One of the focal points of the exhibition was the Regional Centre. This indicated a liking for regionalism – a foretaste of Marshal Pétain's politics – whereby the region was regarded as a protective microcosm in the face of a rising tide of nationalism and imperialism. Furthermore the design of the Craft Centre was indicative of an anti-industrial ideology. The 22 foreign pavilions and indeed the whole exhibition were characterized by a classical modernity which combined the visual arts: for example, large mural frescoes and sculptures were brought together. Much effort was directed towards science and technology. A new sense of modernity was developing – a rational architectural style based on a minimal aesthetic sense. The Bauhaus spirit was making itself felt.

Contemporary design (1939-58)

- Made in the USA
- Japan: the development of 'Good Design'
- Great Britain: from war to peace
- Scandinavia: tradition and innovation
- Germany: starting from scratch
- Post-war Italy: 1948-58
- France discovers industrial design

Human conflicts, being periods of turmoil and experimentation, may lead to the emergence of new materials and manufacturing processes. World War II armaments were remarkable in that aesthetics played a part in their development. After the Liberation of Europe the American model was followed: this was notable for its use of prefabrication in architecture, the prevalence of plastic in household equipment and the flexibility of forms made possible by plywood. For at least a decade, the USA was to impose the American way of life on Europe, its spread assisted by Marshall Aid. A gradual transformation of Europe and Japan got under way, involving a slow readjustment to freedom. The reconstruction of towns and housing was among the foremost of people's concerns and required huge amounts of energy.

Made in the USA

In the USA, designers had been working for large-scale businesses since 1938, the date when the Art and Colour Section was set up at General Motors. The United Society of Industrial Design, the first professional design organization, was founded in New York in 1944.

Public recognition of the profession of designer

The theme of the 1939 International Exhibition in New York was 'building the world of tomorrow'. Designers were ideally placed to respond to this aim. Walter Dorwin Teague, who was on the exhibition's organizing committee, designed pavilions for Ford, American Steel, Eastman Kodak and National Cash Registers. Henry Dreyfuss was responsible for the interior of the 'Perisphere' pavilion, which included *The House of Tomorrow*. Designers were without doubt the key figures at the exhibition, and their profession received public recognition for its role as an interpreter of industry: Marcello Nizzoli, Max Bill, Henry Dreyfuss, Raymond Loewy, Walter Dorwin Teague, Hans Gugelot, Gio Ponti – all the pioneers of international design were there. It was in the USA that the profession of industrial designer became 'official', with the introduction of service contracts between companies and design agencies.

During the 1940s and the early 1950s, the Museum of Modern Art in New York helped to promote design. It organized the 'Low Cost Furniture Design Competition' exhibition (1940–41) in order to support young designers. The initiative came too early: the war delayed its impact. At the same time, the armed conflict brought about major technical advances, as the tools and materials of the war industry were adapted to the furniture industry and the production of household objects. Designers favoured a functionalist aesthetic approach dictated by the requirements of military production and industrial design: 'good design'. Edgar Kaufmann's book *What is Modern Design?* (1950)

Previous page: *Ant chair 3100*, made from a single piece of moulded plywood, Arne Jacobsen, 1952. Manufactured by Fritz Hansen, Denmark.

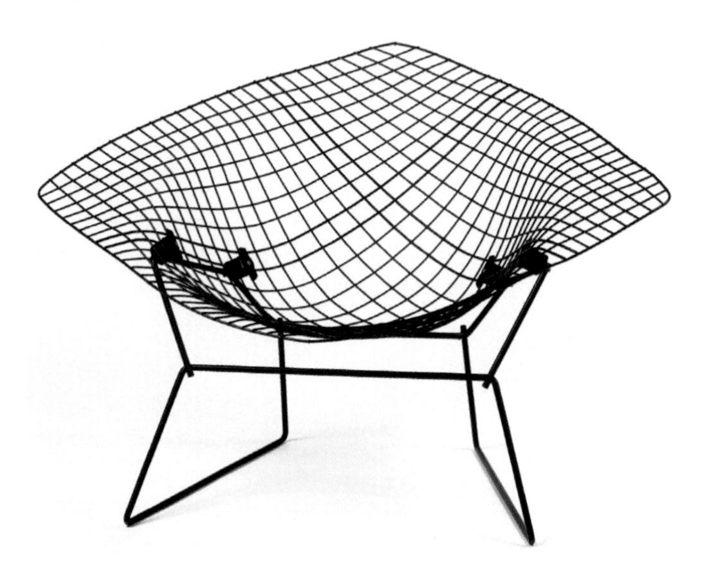

Diamond armchair, with a seat made from a lattice of steel wires bent by hand, Harry Bertoia, 1952. Manufactured by Knoll Associates Inc., New York. Vitra Design Museum.

became the bible of good taste. However, it represented only one aspect of popular taste. Between 1945 and 1960, design logic oscillated between rationalism and organic design.

The consumer society was highly developed in the USA, and the Americans displayed contradictory desires: 'good design' was intended for them, but they found it too austere, preferring the extravagant style favoured by the influential designer Harley J Earl. Working for General Motors, he produced bumpers shaped like torpedoes, with projecting fins; the dashboard was covered with dials. His refrigerators were decorated with hub caps. The control panels on his washing machines looked like dashboards, again inspired by the aesthetics of the car.

A generation influenced by others

The influence of designers from the Bauhaus became more widespread in the field of industrial design. A new generation emerged from the Cranbrook Academy of Art (Bloomfield Hills, Michigan), including Charles Eames, Eero Saarinen and George Nelson. This generation of designers, many of whom had trained as engineers, had a logical, rational professional approach. Their ideas about form were closely related to the function of the object in question. Some – Charles Eames and Henry Dreyfuss, for example – had worked on projects for the army during World War II. The war brought major innovations with regard to materials, techniques and industrialization, and these developments influenced forms. The 'Organic Design in Home Furnishings' exhibition

(1940) stimulated Saarinen's and Eames's imagination. New materials enabled them to come up with softer lines.

Plastics were used in the aeronautical industry (for cockpits in planes), and Plexiglas and fibreglass were used in naval vessels. The adaptation of plastics for domestic purposes was achieved by Earl Tupper, who designed cheap food containers made of polyethylene with airtight lids (Tupperware). Plexiglas (1936) appealed to designers who wanted to show the internal structure of a product. The use of foam rubber or moulded rubber also had a major influence on the development of chairs, providing improved comfort especially for the seat. Eero Saarinen's *Womb Chair* (1948) had a curved form with a moulded plastic shell, fabric-covered latex-foam upholstery and a metal leg-frame.

Developments in plywood resulted from wartime aeronautical research, in which Charles Eames participated. Thin strips of wood, glued together with the grain running in alternate directions, made an extremely tough material capable of being formed into free curves and less breakable than bentwood.

They turned to other materials as well. George Nelson used brushed aluminium for round electric clocks (1955), while Walter Dorwin Teague chose glass for a barometer (1949). On the *Hallicraft* radio (1947), Raymond Loewy replaced the gold colour of the switches with black and white, making it look more like a piece of aeronautical equipment.

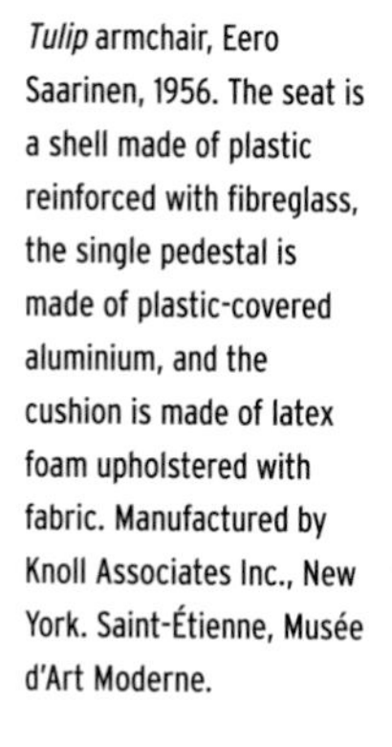

Tulip armchair, Eero Saarinen, 1956. The seat is a shell made of plastic reinforced with fibreglass, the single pedestal is made of plastic-covered aluminium, and the cushion is made of latex foam upholstered with fabric. Manufactured by Knoll Associates Inc., New York. Saint-Étienne, Musée d'Art Moderne.

Charles Eames

Stacking and linking fibreglass *DSS* chair, Charles Eames, 1948. A light, comfortable back-up chair that met a growing demand for office furniture. Manufactured by Herman Miller Inc., USA. Vitra Design Museum.

An architect and a designer, Eames (1907–78) was head of the Design Department at the Cranbrook Academy of Art. In 1940, he and Eero Saarinen won first prize in the 'Organic Design in Home Furnishings' competition held by the Museum of Modern Art in New York with an innovative chair made from preformed plywood. The plywood was shaped in two planes: the upper part of the chair and the seat are made of a single piece, forming a shell. This highly technical project marked the start of a new direction for furniture, but it was not yet intended for mass production. Charles Eames and his wife Ray moved to Los Angeles, where they continued to pursue their technical research into the moulding of plywood. Following a commission from the American navy, they developed splints and stretchers made of preformed plywood which were produced by the Evans Products Company of Los Angeles (1943). Between 1942 and 1946, they designed a variety of chairs made from moulded plywood, such as the *DCW* and *DCM* chairs (1946) for the Evans Products Company. The Herman Miller furniture company took Charles Eames on as a design consultant in 1946, then entered into an exclusive contract with him. That same year a one-man exhibition was devoted to his work at the Museum of Modern Art in New York. Charles Eames was more concerned with the technical aspect of design and with research into materials than with the aesthetic aspect. He designed a chair with a fibreglass-reinforced polyester shell as the seat, which won him second prize in the seating category of the 'Low Cost Furniture' competition (1948), and an armchair made of welded steel wires, the *Wire* chair (1951). His partnership with Herman Miller continued with the *Aluminium Group* (1958), an extremely sophisticated collection of furniture.

A room fitted with a 'storage wall', adjustable shelving and storage units (1949), decorated with the *Ball Wall* clock (1947), George Nelson. The chairs are designed by Charles Eames: *DCM* (Dining Chair Metal, 1945–6), moulded plywood. Manufactured by Herman Miller Inc., USA.

An American way of life

The USA became the major post-war player in design due to several factors: the emigration of European artists, the strength of its economy, the vitality of its industry and its leading-edge technology. American manufacturers attached more importance to the profitability of the object than to the object itself. Production techniques were constantly being improved. Technological advances speeded up the pace at which domestic equipment was replaced, especially towards the end of the 1950s in the fields of electrical goods and household appliances. Given these rapid technological developments, it was now cheaper to buy new than to have an item repaired. Design in the USA was the three-dimensional visual expression of a material culture. Designers who had started off as creators now became coordinators, taking account of consumers' motives for buying and calling on the skills of specialists in economics, marketing and the humanities.

The USA was represented by two large and powerful furniture companies, Herman Miller and Knoll, run by teachers and graduates from the prestigious Cranbrook Academy of Art: Charles Eames, Ray Kaiser (Eames), Eero Saarinen, Harry Bertoia, Florence Schust (who married Hans Knoll) and Maija Grotell. These two companies were a hothouse for the creativity of designers.

Japan: the development of 'Good Design'

During the 15 years that followed World War II, Japan recovered from its devastation to become one of the great industrial nations, manufacturing a large number of advanced industrial products: motorbikes, cameras and television sets, to name but a few.

Design, a national cause

The history of design in Japan is linked not only to this resurgence in manufacture but also to the post-war occupation of the country by the Americans. The Directorate of Japanese Applied Arts was asked by the Americans to fit out lodgings for the occupying troops. Mitsubishi and Toshiba took responsibility for manufacturing the household appliances, and this gave them the opportunity to become acquainted with American industrial technology and the American way of life. An exhibition entitled 'Learning from America: the Art of Daily Living' was put on in Tokyo in 1948. The Korean War helped bring about reconciliation between the USA and Japan. In 1951, the designer Raymond Loewy made a contribution by designing the packet for Peace cigarettes. Through the MITI (Ministry for International Trade and Industry) and the Ministry for Finance the government stimulated growth by promoting the importation and use of foreign technology. The MITI steered design both towards exports and towards the creation of a domestic market,

Herman Miller

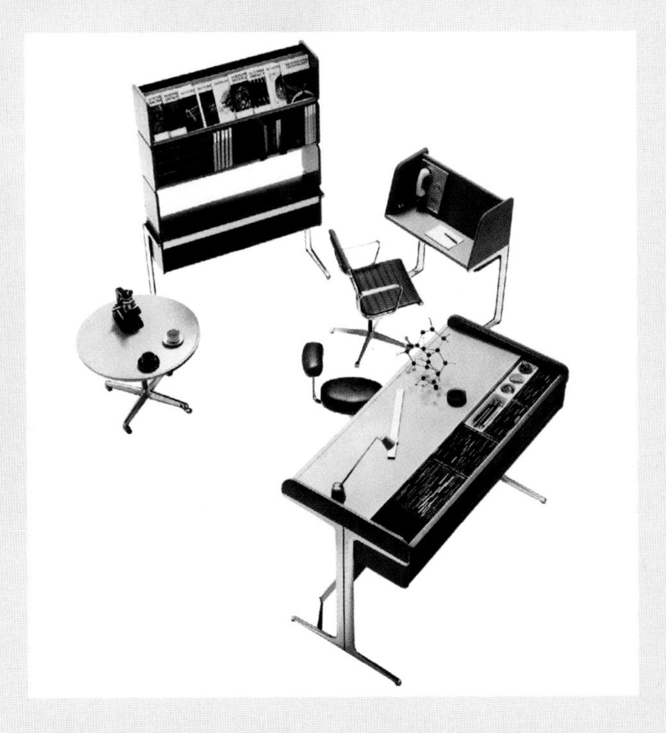

Action Office 1, based on a concept by Robert Propst, George Nelson, 1964. Manufactured by Herman Miller Inc., USA.

The furniture company Herman Miller is based at Zeeland, Michigan. It developed from a small business known as the Star Furniture Company, created in 1905, which produced domestic furniture. In 1923, the company (run by D J De Pree) shifted its focus to contemporary furniture, changing its name to Herman Miller (De Pree's father-in-law) when Gilbert Rohde (1894–1944) joined it in 1931. The emphasis was on modernist design. Both men were motivated by 'honest design', imbued with a kind of morality. In 1942, Rohde designed the *Executive Office Group*, a suite of office furniture. He died in 1944 and was succeeded in 1946 by the architect George Nelson (1907–86).

Together with Henry Wright, Nelson had written *Tomorrow's House* (1945), in which he laid down new principles for the management of living space, in particular advocating compartmentalization by means of storage units. In 1946, he introduced Eames to De Pree. Nelson designed rational storage systems: the *Storage Wall* (1946), the *Steel Frame Group* (1954) with John Pile, *Modular Seating* (1955) and the *CSS Storage System* (1959). *Action Office 1* (1964) is based on the concepts of Robert Propst, whose ideas influenced the business in their development of a complete office furniture programme, including modular storage units, chairs, and so on – a total office environment, further developed in *Action Office 2* (1968). Alexander Girard (1907–93) joined the team in 1952. Appointed director of the textiles department, he introduced wallpapers and fabrics in sumptuous colours and new patterns. The imperatives of their design policy were aesthetic as well as technological and functional. In the Herman Miller design process, designers worked in close collaboration with the factory, and a large part of the work was still done by hand.

IBM (International Business Machines Corporation)

IBM *Selectric* typewriter, Eliot Noyes, 1961. The golf-ball system was a major technological innovation. Noyes was responsible for the overall image of the IBM company. Paris, CCI-Georges Pompidou Centre/Kandinsky Library.

IBM – an American information technology company that Thomas J Watson created in 1914 by merging several smaller businesses – was given its present name in 1924. Due to an excellent sales force and the quality it offered, it expanded considerably. In 1955, Thomas J Watson Jr, who would become chairman of the company the following year, stated his conviction that 'Good design is good business'. He launched a programme centred on the computer and brought in Eliot Noyes (1910–77), who was design consultant from 1956 to 1977. An architect by training, he had worked for the former Bauhaus teachers Marcel Breuer and Walter Gropius. He had been director of design for Norman Bel Geddes, before opening his own design agency in 1947. Of all the freelance designers who embarked on their careers after the war, Noyes was the one most warmly received by the critics. He ran the Department of Design at the Museum of Modern Art in New York for four years. He was the designer in charge of the Ramac computer programme. The *Ramac 305* was launched on the international market in 1957, a remarkable example of a product governed first and foremost by functionalist rules. For IBM, Noyes designed the *Executive* typewriter (1959), the *Selectric* typewriter (1961), the *Executary* dictating machine (1961) and the *1440* computer (1962). The company brought out the first completely transistorized computer in 1959. IBM made a crucial improvement to the typewriter in 1961, when they introduced the golf-ball system in the *Selectric*.
Paul Rand, who was responsible for the company's graphic communication department, also made a major contribution to IBM's success.

and in 1957 introduced a range of 'good design' products which were awarded the 'G-mark', thus encouraging innovation. The organization running the scheme was the Council for the Promotion of Design.

From 1950 onwards, the Japanese followed the American model, opening design agencies which operated on functionalist, rationalist lines. Design was a modern profession, and for the first time manufacturers employed designers to improve their products. Jiro Kusogi started work with Tokyo Kogyo (which became Mazda) in 1949, Sori Yanagi worked for the furniture manufacturer Tendo Mokko and for the porcelain factory of the Tajimi Institute in 1956, and Yusaku Kamekura joined the Nippon Kogaku company (which became Nikon) in 1954. In response to the need for recognition and professionalism, several professional associations were created, including the Japan Industrial Designers Association (JIDA). Isamu Kenmochi was one of the first Japanese designers to travel to Europe and the USA: he visited the 'good design' exhibitions at the Museum of Modern Art in New York, drawing inspiration from them. Together with Yanagi, Riki Watanabe, Kamekura, the architect Kenzo Tange and a few others, he founded the Committee of International Design in 1953, renamed the Committee of Good Design in 1959. They selected products and exhibited them regularly in the 'Good Design Corner' of the Matsuya department store, thereby raising the Japanese public's awareness of the influence of design in domestic and industrial life.

Sony, a model company

In 1953, the Sony Corporation bought from the American Western Electric Company the licence for a process which it then developed to produce the first Japanese transistor radio. When it was set up in 1945, the company was called Tokyo Tsuchin Kogyo. It produced its first tape recorders in 1950. In 1955, the first successful transistor radio, the *TR-55* (known as the *Sony*), using American patents, was a resounding success. The name of that model was adopted as the company name in 1958. That same year, Sony was awarded a gold medal at the World Fair in Brussels, and in 1960 it received another at the Milan Triennale. Sony developed transistorized television in 1959, and the *SV-201* transistorized video recorder in 1961. The company was innovative in the field of miniaturizing equipment and ran a high-quality research programme, but competition was keen and Mitsubishi, Hitachi and Matsushita were fierce rivals. Internationally, Sony was regarded as a state-of-the-art company.

Tradition and looking to the future

Japanese design was undoubtedly the first post-modern style. It actively exploited its history and Japanese traditions, while at the same time designers created completely innovative objects. Zenichi Mano, for

TR-55 Sony radio, 1955. This was the first completely transistorized radio, mass-produced by the Tokyo Tsuchin Kogyo company. In 1958, the company adopted the name Sony.

example, drew inspiration from the shutters on 17th-century buildings in designing the front of a radio set; he won a prize awarded by the *Mainichi* newspaper, another organ promoting industrial design. The manufacture of products such as packaging was a continuation of the age-old tradition of origami, or folded paper. By developing elements from traditional culture and handicraft techniques and combining them with modern technology, Japanese design acquired durable, identifiable qualities.

Japan became a global economic power. It consolidated its position as leader in certain sectors, especially miniaturization, as demonstrated by the *Subaru 360* car (1958), and first and foremost (and very rapidly) electronics. The Sony Corporation was at the forefront of developments, ahead of the Honda Motor Company, Nikon and Fuji.

In 1969, the National Museum of Modern Art in Kyoto staged the 'Exhibition of Modern Design in Japan'. It featured a wide selection of Japanese graphic design and furniture, with pieces by Saburo Inui, Kenzo Tange, Bunsho Yamaguchi, Daisaku Cho and Munemichi Yanagi, all produced by the Tendo Mokko company. There were also designs by Isamu Kenmochi for the company Yamakawa Raltan, a lamp by Isamu Noguchi for Akari, handicrafts (fabrics and ceramics), and examples of industrial design: a Canon cine-camera, a Minolta projector, a Seiko chronometer, a Sony transistor radio and portable television, and a Pioneer amplifier. This exhibition demonstrated the extraordinary creative dynamism of Japan, now a major industrial nation on which European and American eyes were focused.

Great Britain: from war to peace

The strategy for winning the war was worked out in Britain. This required the mobilization of industry and the concentration of the entire nation's efforts. The country endured a great many attacks from the direction of continental Europe, with the south of England suffering enormous damage. The government organized industry in such a way that it would serve the war effort and help the civilian population.

Utility products

In 1942, Great Britain set up the Utility Furniture Advisory Committee, created by Hugh Dalton in his role as President of the Board of Trade. He masterminded a strict policy governed by rationing restrictions, and appointed the furniture manufacturer Gordon Russell (1892–1980) to the committee, eventually making him head of the Design Panel. Both men adhered to the principle that businesses would be allowed to produce only a limited range of products approved by the Committee. This restriction applied to clothing, china and furniture, and guaranteed simple, good-quality products made from decent materials. The plan proved hard to implement. Russell ran the Design Panel with the help of Enid Marx, who was put in charge of furnishing fabrics from 1944 to 1947. She favoured a restricted range of colours and small repeat patterns to prevent the wastage of raw materials. The committee laid down criteria of simplicity and effectiveness which were in fact the criteria of modernity. However, rules were applied more flexibly as the end of the war drew near and the British people grew tired of these austere products. The committee turned towards smaller, more sophisticated pieces of furniture which were praised in *The Architectural Review*. The DRU (Design Research Unit) was founded in 1943, with Misha Black and Milner Gray as the leading designers. The DRU designed stands for the 'Britain Can Make It' exhibition (1946) and the Festival of Britain (1951). It carried out interior decoration and architectural schemes as well as industrial design projects, designed product lines and acted as consultants in all relevant sectors.

Britain Can Make It

The Council of Industrial Design (CoID) was founded in 1944 as part of the reconstruction policy designed to promote the proper use of design and heighten public awareness of it. Under the leadership of Gordon Russell, the Council helped the Design Centres (offshoots of the CoID) in London and Glasgow to stage exhibitions. 'Britain Can Make It', held at the Victoria and Albert Museum in London in 1946, marked a turning point in Britain. Officially opened by King George VI, the event attracted one and a half million visitors. In the 'War to Peace' section, Britons could observe the impact that military research had made on everyday objects. Ernest Race produced the *BA* chair (1945): its frame was made from recycled aluminium alloy recovered from warplane scrap. It was put on the market, with 250,000 chairs being produced. The success of the exhibition inspired a series of other events which contributed towards a wider recognition of British contemporary style. 'Design at Work' was a smaller exhibition relating to British industry (1948). Nonetheless, it showed products made by the Hille furniture company and the work of the young designers Robin Day and Clive Latimer, who that same year won first prize in the international 'Low Cost Furniture Design' competition organized by the Museum of Modern Art in New York.

Restrictions continued, with shortages persisting until 1950, when Marshall Aid was implemented. In 1951, the Institute of Contemporary Arts (ICA) staged an exhibition of experimental furniture, including designs by Geoffrey Dunn. In 1949, the Council of Industrial Design launched *Design* magazine, with the remit to spread the message of modernity. Robin Darwin, the principal of the Royal College of Art from 1948 to 1971, the sculptor Eduardo Paolozzi, who taught textiles at the Central School of Arts and Crafts (1950–55), and Robin Welch, a teacher at the Central School of Art and Design (1957–63), were all educationalists who had a lasting influence. Herbert Read made a crucial contribution with his book *Education through Art* (1942). *House and Garden*, a magazine on decorating the home, supported the movement by publicizing the ColD's selected products as well as goods made by Heal and Son.

Under the stewardship of Gordon Russell, who led vigorous campaigns to heighten public awareness, as well as through exhibitions, prospectuses and lectures, the Council of Industrial Design began, around 1950, to reap the rewards of its strenuous efforts. Industrial

BA chair, frame made from recycled aluminium alloy recovered from warplane scrap, Ernest Race, 1945. Manufactured by Race Furniture Limited. Some 250,000 chairs had been produced by the end of the 1940s.

design in Britain improved thanks to the objects produced by companies such as Hille, Race Furniture Ltd and Morris of Glasgow. The notion of design had finally come to be accepted as normal. After all the years of adjustment or obstinacy, the Festival of Britain (1951) – a major international exhibition decided upon in 1947 as a centenary celebration of the Great Exhibition in London – represented a breakthrough. In 1949, the project was scaled down and turned into a national festival, in order to stimulate national feeling and promote the idea of 'progress'. The Council of Industrial Design was heavily involved in the project and made a selection of 10,000 manufactured objects. That documented selection became the *Design Review*. A detailed document was created for each product (whether hen house, railway engine, cutlery or sailing boat), and objects were divided into 70 categories. Many designers contributed towards preparing the Festival of Britain pavilions: Robert Goodden and Dick Russell (the Lion and the Unicorn pavilion); Katz & Vaughan (Homes and Gardens); Welles Coates (the Telecinema); Jack Howe (rubbish bins); Ernest Race (chairs); and Milner Gray and Robin Day (signs). The exhibition provided a number of designers with their first commissions. Race was asked to produce his famous *Springbok* and *Antelope* chairs. Day was commissioned to design all the seating for the Festival Hall, and many young artists were engaged to paint frescoes in the same venue. A new style was emerging in Britain.

International success

A new type of designer appeared on the scene: they were neither architects, artists, nor craftsmen. These ambitious young Britons preferred the Scandinavian idea of independent design bureaux to the American system of large agencies. On leaving the Royal College of Art, they opened their own agencies, motivated by a strong feeling for detail, self-belief and a love of work well done. Royal College of Art graduates included the furniture designers Ronald Carter and Robert Heritage, the textile designers Audrey Levy and Pat Albeck, the jewellery designer John Donald, and the industrial designers Gerald Benney, Robert Welch and David Mellor. The Council of Industrial Design was rewarded for its efforts by the long-awaited opening of the Design Council Centre in 1956, with the backing of industry and Gordon Russell, among others. In 1957, the prizes in a competition organized by the Design Council Centre were awarded for the first time to a partnership of designer and industrialist: Robin Day and Hille, the Reids and Rotaflex, Hulme Chadwick and Wilkinson Sword, Robert Heritage and Archie Shine and the Ogle Design company (1954), Conran Design (1955) and Leslie Gooday & Associates. One of the first schemes to be implemented with no restrictive parameters – the constraints imposed by the post-war period were at an end – was the building and furnishing of the Time Life Building by Michael Rosenauer in London. The creation of new towns such as Basildon, Harlow and Hatfield, with their innovative architecture, necessitated public amenities, suitable furniture, schools, day-care centres, and appropriate parks and gardens. Furthermore, the baby boom tended to enhance the creative dynamism of design even more.

Management and consultants

From 1957 onwards, design was transformed into a veritable means of seduction for the British, by now fully immersed in the consumer society. Businesses (airline companies, banks, and so on) called on designers to revamp their image and provide them with a visual identity. Design acted as advertising. It was a sales asset and as such of direct concern to business management, and for the first time businessmen who invested in design were rewarded commercially and economically. The British style became closer to the international style. In just a few years, the number of magazines focusing on the home increased dramatically: *Home, Homes and Gardens, House Beautiful, Housewife, Everywoman, Woman and Home* and *Woman's Journal.*

Coming full circle

At this period, the aesthetic sense of the Scandinavian countries was very much admired and regarded throughout the world as representative of good taste. Paul Stemann, on behalf of the company Finmar, imported chairs by Hans Wegner, ceramics, gold work, silverware and lamps by Kaare Klint. The shop Green Group also started to import a large number of Scandinavian articles. The first shop to do so was Primavera, Henry Rothschild's boutique on London's Sloane Street, in 1945. Many other sales outlets opened throughout Britain. Scandinavia represented the ultimate in design: simplicity, beauty of natural materials, visibility of the construction process and comfort. Design had come full circle. Denmark, influenced by the 19th-century Arts and Crafts movement, was in turn influencing English design in the 1950s.

Scandinavia: tradition and innovation

The Scandinavian countries played a major role in the field of design. The fact that they were rooted in the tradition of trades relating to wood guaranteed them great stability in the years ahead.

A Scandinavian identity

Styles and references which made an impact at the Milan Triennale, first held in 1923, had a decisive influence on design. The Triennale made a major contribution to the international recognition of Scandinavian design. During the 1940s, only Sweden, which had not been affected by the war because of its neutrality, was able to exhibit. In 1948, it won two gold medals, awarded to Berndt Friberg and Stig Lindberg, ceramicists

PK24 hammock chair, steel, cane and leather, Poul Kjaerholm, 1965. It follows the shape of the body; the position of the leather headrest cushion can be adjusted. Manufactured by E Kold Christensen. Vitra Design Museum.

at the Gustavsberg factory. After the war, the other Scandinavian countries also started to win prizes. Scandinavia dominated the Triennale for 20 years. As early as 1951, four 'Grand Prix' awards were made to Scandinavian countries: two to the Finns (Tapio Wirkkala for the Finnish pavilion with its amazing glasswork and Dora Jung for her textiles) and two to the Danes (Hans Wegner for his furniture and Kay Bojesen for his silverware). A silver medal was awarded to the Finnish glass-maker Timo Sarpaneva. The same result was achieved in 1954, and in 1957 there were 'Grand Prix' awards for the Danes Arne Jacobsen and Poul Kjaerholm, and the Finns Sarpaneva and Kaj Franck. In 1960, apart from the 'Grand Prix' award for Wirkkala and Kjaerholm, there was a gold medal for Ole Wanscher and Nanna Ditzel. These exhibitions had major repercussions on sales, and the USA became the first customer for Scandinavian design. In 1954, the 'Design in Scandinavia' exhibition embarked on a three-year tour of the USA, ensuring widespread publicity for Scandinavian design. Other exhibitions conjured up the special spirit that characterized the Nordic countries. The most important of these, entitled 'H55', was held in Hälsingborg in southern Sweden in 1955.

Danish supremacy

Danish design at its height is represented by Hans Wegner (b.1914). He collaborated with Arne Jacobsen and opened his own practice in 1943. He designed many chairs and exhibited regularly at the Guild of Cabinetmakers.

His model *JH501*, known as the *Round Chair* (1949), was hugely successful and made the front covers of American magazines. Arne Jacobsen (1902–71) approached design from the point of view of an architect. For the Royal Hotel in Copenhagen he designed the *Egg*, *Swan*, *Drop* and *Pot* chairs, revolutionary in their forms and materials: polyurethane foam covered with leather and fabric. He also designed the lights and tableware (1959). Created for the Novo pharmaceutical laboratory, the *Ant* chair (1952) manufactured by Fritz Hansen was immensely successful. Poul Henningsen (1894–1967) carried out research into electricity and designed many lamps. In 1958, he came close to perfection with the *Kogle* chandelier (also known as the *Artichoke* lamp), manufactured by Louis Poulsen. Poul Kjaerholm (1929–80), a prominent figure in Danish design, used flat steel in conjunction with leather and plaited cane. His hammock chair *PK24* (1965), his folding stool *PK41* (1961) and his hammock chair *PK45* (1965) are remarkable for their simplicity. All his furniture was produced by the firm of Kold Christensen.

Sweden – a head start

Having remained neutral during the war, Sweden got off to a head start. In the post-war years, the country was destabilized by a rural exodus. In order to accommodate the new city-dwellers, a major infrastructure programme was implemented. This was a good period for designers and for

Ant chair 3100, made from a single piece of moulded plywood, steel legs, Arne Jacobsen, 1952. Manufactured by Fritz Hansen, Denmark. Saint-Étienne, Musée d'Art Moderne.

Ikea

Folding chair made of tubular metal and canvas, manufactured by Ikea.

Ingvar Kamprad, the founder of Ikea, introduced a new businesss idea in 1943: to manufacture light furniture in large production runs. A catalogue of the collection was published from 1951 onwards. The first large Ikea home furnishing store opened in Almhult in 1953. The company employed its own designers: the 1950s furniture was designed by Bengt Ruda and Erik Worts. They were pioneers of Ikea design, producing furniture made of oiled teak, woven curtains and the *Ogla* chair.

In 1956, furniture became available in flat-pack kit form. The furniture was designed to allow for maximum standardization. Elias Svedberg and Lena Larsson created *Triva*, a series of furniture items in kit form. In 1965, Ikea stores opened in Stockholm. Their sales policy involved one major novelty. The furniture was not sold in traditional shops, but in furnishing superstores which stocked everything relating to interior furnishing and its accessories: lighting, carpets, glassware, fabrics, china and kitchen utensils. Furniture was sold on a self-service basis. These warehouse-cum-stores, located outside towns in industrial zones where rents were relatively low, could be easily reached by car, enabling buyers to take their goods home with them without the need for the store to deliver. The furniture was supplied with assembly instructions. All these factors helped to reduce the cost of the furniture. The Möbelfakta label, based on the standards of the Swedish Furniture Institute, became part of Ikea's internal quality control. The Ikea brand was launched onto the international market in 1973.

industries related to the visual arts, with close collaboration between furniture, textile, ceramics and glass companies.

Following Ikea's example, the cooperative chain stores KF Interiors and Domus embarked on a policy of expansion. KF had been the leading business for cheap, good-quality furniture since the 1940s. In collaboration with the Swedish Society of Industrial Design, it had introduced techniques designed to improve the quality of furniture. There were many research projects looking at the quality and durability of furniture, including various tests and trials. The results were passed on to consumers and legally certified.

In the 1950s, Nils Strinning introduced a new type of shelving, which involved laying wooden boards on metal brackets. Yngve Ekström designed the *Domino* furniture collection, which enjoyed tremendous success in government departments and offices. Craftsman-like style, careful thought about the industrial aspect, standard elements used on the modular principle – all these made the production of high-quality objects possible.

Finland

After the war, the designs of Pirkko Stenros and Saara Hopea made a great impact. In the 1950s, when the notion of Scandinavian design was developing, Finland stood out because of its exceptional innovative achievements in the applied arts: the *Chanterelle* vase (1947) by Tapio Wirkkala has truly iconic status. Timo Sarpaneva and Antti and Vuokko Nurmesniemi were the leading figures in Finnish design.

Germany: starting from scratch

Post-war Germany moved away from the tradition of arts and crafts embodied in *Kunsthandwerk*, which had been promoted by National Socialism, and opened up to the dynamics of technology. From 1949 onwards, it very quickly became a highly industrialized country again, focusing on modernity. The watchword was 'functionalism'.

Rat für Formgebung

The Rat für Formgebung (Design Council) was founded by the German government in 1951. It was financed by central state funds, funds from the *Länder* (federal states) and private donations. Based in Darmstadt, this body played a role in the reconstruction and repositioning of Germany after World War II. At the time it was founded, German industrial design was virtually non-existent. The work of the pre-war pioneers had been forgotten. However, the Rat für Formgebung initiative shared some of the intentions of the German Werkbund, which had been dissolved in 1933. The objective of the Rat für Formgebung was to ensure that German industrial and craft products were as well designed as possible and therefore more acceptable to the consumer. In this way, design played both a cultural role and an economic role in encouraging consumerism. The articles of association of the Rat für Formgebung, revised in 1964, specified its objectives: research into design, education through design, and the coordination of cooperation between local authorities, industrialists, businessmen and designers. In particular, the organization arranged contacts between businesses and designers by publishing the *Deutsche Warenkunde*, an illustrated publication devoted to German consumer products. The Rat für Formgebung encouraged German designers to take part in as many exhibitions as possible: the Milan Triennale (1954, 1957, 1960 and 1964), the Interbau in Berlin (1957) and the Brussels World Fair (1958). In 1965, the 'Gute Form' exhibition at the Council of Industrial Design in London, mainly devoted to technical design, presented a selection of high-quality German products. Most of the companies represented – AEG, Bosch, Grundig, Olympia, Pfaff, Siemens and Zeiss – had their own in-house design agency.

Ulm School of Design

1947–53. Inge Scholl, Otl Aicher, Max Bill and the Austrian Walter

Max Bill, who had studied at the Bauhaus, was the first director of the Ulm School of Design, from 1953 to 1957. He was also the architect of the buildings of the School, which were officially opened in 1955.

Zeischegg, as well as many others including the writers in the 47 Group, came together to define their project for an educational institution and organize the ways and means of setting it up. The Americans lent their support, and in 1949 the group successfully staged the 'Gute Form' exhibition in Ulm – an exhibition which set out their idealistic aspirations. At a time when Germany was still a country in ruins, they decided to promote the teaching of design.

The Ulm School of Design (Hochschule für Gestaltung Ulm) was the most important design institution and the standard-bearer of the discipline. The new school owed its existence to a private foundation in the name of Hans and Sophie Scholl, young resistance fighters executed by the Nazis, and this guaranteed its intellectual independence – necessary for defending a new experimental approach and a critical attitude to society, especially when it had to deal with a traditional and conservative bureaucracy. Other Germans, such as the physicist Werner Heisenberg and the writer Carl Zuckmayer, were also looking for change and renewal and they encouraged the country to move forward.

1955–62. The School was officially opened in 1955. Max Bill (1908–94), who had studied at the Bauhaus, became its first director. He designed the building, an embodiment of his ambition: 'From the spoon to the town ... collaborating in the construction of a new civilization.' He secured the services of a very prestigious international teaching team with new ways of thinking, including the Argentinian painter Tomás Maldonado, the Dutch architect Hans Gugelot (1920–65) and Friedrich Vordemberghe-Gildewart, a former member of the De Stijl group. While Bill's principles were those of an artist, the teaching-manufacturing concept advocated by Gugelot was introduced from 1955 with the collaboration of the company Braun, which had approached teachers at the School. The following year Dieter Rams came up with the concept of on an unornamented style for Braun products. Bill's younger colleagues championed an approach geared towards science and modern mass-production technologies. They were thus radically opposed to the Bauhaus theories on handicrafts represented by Bill, and he decided to leave Ulm in 1957.

Maldonado succeeded him in 1958 and the School entered a new phase, influenced by the teaching of Aicher, Gugelot, Zeischegg and Vordemberghe-Gildewart. They tried to establish still closer links between design, science and technology. The designer would have to be more modest: he could no longer regard himself as an artist, but was encouraged to adapt to working in a group including scientists, researchers, businessmen and technicians. An 'Ulm' model was established, once again questioning the ultimate purpose of design. From 1958 to 1962, the need to include the humanities, ergonomics, operational sciences and the methodology of industrial technology planning

TC100 stacking hotel crockery, Hans Roericht, designed when he was studying at Ulm, 1959. The design was put into production by Thomas-Rosenthal AG.

in the courses raised the problem of a growing imbalance between scientists and designers. 'Working hypotheses' replaced manifestos. Another powerful idea that originated in Ulm was the 'design system', formulated and applied by Gugelot. Based on analysing the typology of objects, putting them in order and matching them or establishing their complementarity, the design system arranged products into groups with interchangeable parts. This system was applied to the whole corporate image.

A generally positive assessment. Under the leadership of Aicher and Maldonado (1962–6), the School tried to restore a proper balance between theory and practice, and between science and design. These continual reappraisals and debates undermined its financial strength. In 1968, the provincial parliament in Stuttgart decided that the School should be closed down. Seen historically, the School was an international centre for education and research in the design of objects intended for use in many fields: everyday use, scientific use, tool-making, the tertiary sector, and architecture and construction. The School also helped to develop the public relations sector with its definition of the corporate image and the creation of corporate style guides and logos. The study of social and cultural factors now became a compulsory part of the curriculum for those training to be designers. Some 640 students trained at

Braun

SK4 radiogram, Hans Gugelot and Dieter Rams, 1956. The lid is made of Plexiglas, the case of light wood and painted metal. Manufactured by Braun, Frankfurt. Paris, CCI–Georges Pompidou Centre/Kandinsky Library.

The Braun company was founded in Frankfurt am Main in 1921 by Max Braun, a manufacturer of radio accessories. From 1950 onwards, it began to produce domestic appliances. In 1951, Max Braun's two sons, Artur and Erwin, took over the company. They were determined to change the look of the products. Not having an in-house design agency at that time, they employed outside designers: Fritz Eichler, who got the desired transformation under way; Wilhelm Wagenfeld, who had studied at the Bauhaus; and Hans Gugelot, who was head of the product design department at the Ulm School of Design. This change of direction was not confined to the appearance of the products; it also affected production and public relations (mass production, graphics, packaging, advertising, and so on). Wagenfeld designed the portable radiogram (1955), while Artur Braun and Eichler produced the *SK1/2* radio with a moulded Bakelite frame, a perforated metal grille concealing the loudspeaker, and simplified controls. Gugelot designed many products for Braun: the *TSG* radio (1955), in collaboration with Helmut Müller-Kühn, one of his pupils at the Ulm School of Design; the *PKG1* and *PKG2* radiograms, which were strictly geometrical in form; and the *SK4* radiogram (1956), produced with Dieter Rams, which was still basic in form but used a variety of materials: folded painted sheet aluminium, Plexiglas and varnished beech wood. The architect and decorator Dieter Rams (b.1932) took over as artistic director of the company in 1961, a role he fulfilled until 1995. He remained very close to the teaching of the Ulm School of Design, developing a demanding policy based on quality and functionalist rigour. He designed furniture and shelving systems for Otto Zapf and carried out projects for De Padova in Italy.

the School; 40–50% of them were foreigners. Ulm succeeded in gaining acceptance for the idea that the cultural values of design related not only to housing but to schools, government offices, factories, hospitals and public transport as well. The School played a major role in international design and helped to bring about a change of attitude towards mass-produced industrial objects on the part of consumers. It was instrumental in promoting the development of corporate public relations, encouraging every company to devise a total image for itself. Braun was the best example of this image-making, which was just as emblematic as the Werkbund-influenced work of the designer Peter Behrens for AEG in 1907.

Protestantism and discipline

With full employment, Germany achieved a good standard of living in the late 1950s. Design as championed by the Ulm School had become associated with a culture of Protestant discipline and a Social Democratic vision of society. In 1954, the philosopher Martin Heidegger published a lecture ('The Essence of Technique') in which he stated that it was necessary for ideas to go beyond pure technique. The system of industrial production was growing rapidly and could cope well with products with smooth, clean lines. Many businesses were set up and expanded. Vitra, a furniture manufacturing company, was founded in 1950. In 1955, Siemens brought out Norbert Schlagheck's numberless clock, and BMW licensed the manufacture of the small car *Isetta*. In 1956, Richard Sapper did work for Mercedes-Benz, while Hans Gugelot and Dieter Rams designed the *SK4* radiogram for Braun. In 1957, the international architecture and design exhibition 'Interbau' was held in Berlin. Bringing together architects from every country, the exhibition hosted a competition on the theme of rebuilding the Hansa district in Berlin. Le Corbusier, Walter Gropius, Alvar Aalto and the Berliner Klaus Müller-Rehm took part in it, but their projects proved disappointing, looking more like prototypes than the result of real thinking about a specific district.

Post-war Italy: 1948–58

Italian intellectuals and artists embarked on an in-depth exploration of reality, as demonstrated by neo-realist cinema, realism in philosophy, existentialism, town planning, the planned economy and functionalism in architecture and design. Italy, contrary to the romantic, light-hearted image it often conjures up, proved to be extraordinarily efficient. Along with cinema, Italian design was certainly one of the main phenomena of the 1950s. Through a combination of industrial and cultural strengths, it became a symbol and a model.

A renaissance

The way had been prepared by Mussolini's Italy, which had been anything but a backward-looking country. Mussolini had advocated a rational, modern image that matched his fascist political ambitions: a new architecture for a new society. Two of the architects active under his oppressive regime were Giuseppe Terragni and Agnoldomenico Pica. Interior designers included Luigi Figini, Gino Pollini and the Castiglioni brothers. Franco Albini was one of the active members of a group of rationalist artists backed by the *Casabella* journal. He then moved into interior decoration and industrial design, playing an active part in the country's cultural life from 1936 onwards. In 1938, he designed a free-standing bookcase with glass shelves suspended from bracing wires, followed by a radio set made of glass. A few companies survived and flourished: Campari, Pirelli, Olivetti. Gino Sarfatti founded Arteluce (a company manufacturing lighting) in 1939.

With a need for capital goods after the war, the design studios got back to work. In a country partly destroyed by the war and weighed down by 20 years of fascism, design was one means of relaunching the economy and increasing production – a new tool for a new industry that could really change things. Design had its finger on the pulse of society. One of the first projects was the monument in Milan to the victims of the extermination camps, carried out in 1946 by the studio of BBPR. In 1947, the first Triennale of the post-war period had 'Housing and Mass Production in the Home' as its theme. The Triennale commissioner, Piero Bottoni, was a Communist. He focused on the problems of the less privileged members of society.

Industry on the march

Italian design modelled itself on American design. A new generation took over the reins. These designers, many of them architects, went into partnership with business leaders.

Coffee machine for Pavoni, Gio Ponti, 1949. Gio Ponti was both an architect and a designer.

The engineer Corradino d'Ascanio designed the *Vespa* scooter with its sensuous forms as a commission for Enrico Piaggio in 1946. The scooter was one of the most innovative means of transport in post-war Italy, becoming the symbol of a way of life.

Businesses. When Giulio Castelli founded Kartell in 1949, he was following in the footsteps of his father, a pioneer in research into the use of high-grade plastics. Giulio, however, envisaged the use of plastic throughout the home. He was in contact with a great many artists and architects. In 1957, he was one of the people behind the ADI (Association of Industrial Design). Kartell was awarded the Compasso d'Oro for various objects made of plastic: lids, lemon-squeezers and plate stands. It was the first company to champion Italian design by producing aesthetically pleasing and functional items for everyday use.

The Borsani family furniture business was developed by the brothers Osvaldo (an architect) and Fulgenzio (a salesman). From 1952 to the mid-1960s, all its furniture was designed by Osvaldo. At the 1954 Triennale they renamed the business Tecno, in order to convey an image of modernity and technological research.

Designers. Gio Ponti (1891–1979) was an architect. In 1949, he designed the Pavoni coffee machine and the Visetta sewing machine, and between 1950 and 1952 he produced wall decorations for liners. In 1951, the *Superleggera* chair, produced by the Cesare Cassina furniture company, was described as 'a chair that needs no qualification, the epitome of chair'. This, his first submission to the Triennale, was spectacularly successful. Ponti's contribution to design continued with standard furniture for the Ninth Triennale and prototypes for cutlery, then in 1953 with the *Distex* chaise longue made by Cassina and sanitary equipment for Ideal Standard. In 1957, Christofle presented Ponti's work in the 'Formes Idées d'Italie' exhibition. Then the three Castiglioni brothers (Achille, Pier Giacomo and Livio) appeared on the scene. Passionate about design, they were invited to organize exhibitions incorporating sound, lighting and decoration. They played an active part in the Triennales from 1947 to 1964. With Alberto Rosselli, Roberto Menghi and Marcello Nizzoli they organized the pavilion for the Tenth Triennale, and designed for the companies Brionvega, Flos, Knoll, Kartell and Zanotta. From 1948 onwards, Marco Zanuso explored the use of tubular metal, plywood and foam rubber in furniture-making. In 1951 the famous *Lady* armchair upholstered in foam was the first of a long series of chairs he designed for Arflex. The year 1958 marked the start of his collaboration with Brionvega, which resulted in the *Antares* television, followed by the *TS504* radio in 1964.

Promoting design

Ponti took over as editor of the journal *Domus*, which attracted an international readership. In 1952, *Domus* reported on the extraordinary vitality of Italian design, remarking on the 'Italian line' – beautiful, pure, simple and internationally recognized. In 1954, the La Rinascente department store – already very committed to design in its collaboration with Carlo Pagani and Bruno Munari – set up a competition, the Compasso d'Oro, to encourage the improvement of industrial or craft products on a technical and aesthetic level. The selection of the winning items would be made at the Triennale. The first competition saw awards given to Carlo de Carli's furniture for Cassina, a foam toy by Bruno Munari, Olivetti's *Lettera 22* typewriter, Marcello Nizzoli's sewing machine, and Gino Sarfatti's *Model 55* cylindrical lamp for Arteluce. The Tenth Triennale was expanded to include architecture and industry. These exhibitions showed what research was being carried out in the field of contemporary living space. The Eleventh Triennale in 1957 suggested that there was a correlation between modern decorative arts, industrial arts and modern architecture.

Sensuous curves

This period was characterized by tremendous activity. Handicrafts were rediscovered and traditional materials such as glass (as in the wonderful creations of Fontana Arte and Venini), ceramics or wrought iron were used, although the creative input of Gio Ponti ensured that the final design products were modern. The exhibition pavilions jointly designed by Luciano Baldessari, Lucio Fontana and Ernesto Breda at the Milan Trade Fairs in 1953 and 1954, as well as the organization of the Ninth Milan Triennale (1951), marked the arrival of Italian design on the international stage. In the field of furniture too, a few outstanding designers emerged: Gio Ponti with his furniture decorated by Piero Fornasetti,

Olivetti

Lettera 22 portable typewriter, Marcello Nizzoli for Olivetti, 1950. Paris, CCI-Georges Pompidou Centre/Kandinsky Library.

Camilo Olivetti founded the business in 1908. In 1912, he declared that a typewriter should be a tasteful, decorative element in a sitting-room. In 1928, the first public relations department was set up by Adriano Olivetti, and the painter Alexander Schawinsky, who had worked with Kandinsky, Moholy-Nagy and Herbert Bayer at the Bauhaus from 1924 to 1928, was put in charge. The department grew with the arrival of Albini, Figini and Pollini in 1931. In 1937, Figini and Pollini, who were architects, designed workers' housing for the company. In 1940, Adriano Olivetti was awarded the gold medal in the graphic arts section at the Triennale for the magazine *Tecnica e Organizzazione*. In 1946, a department of advertising techniques was set up, headed by Giovanni Pintori. The Olivetti Cultural Centre opened in Ivrea in 1950, with an exhibition devoted to '25 years of Italian Painting'. At this time the factory was producing typewriters. Many publications referred to Olivetti's dynamism, from the monthly internal newsletters *Notizie Olivetti* to the numerous books on the company's history, such as *Twenty-Five Years of Olivetti* (1933) or Schawinsky's *The History of Writing* (1938). The *Lexicon 80* typewriter and the *Divisumma* calculator were launched in 1948, designed by Giuseppe Beccio and Natale Cappellaro with the collaboration of Nizzoli. The *Lettera 22*, Olivetti's first portable typewriter, appeared in 1952. The business expanded and offices sprang up in cities throughout the world, including New York (1950), São Paulo and Sydney (1952) and Frankfurt (1953). Many showrooms fitted out by major designers opened at prestigious addresses: Milan (Nizzoli), Paris (Albini), New York (Belgiojoso Peressutti and Rogers) and Venice (Carlo Scarpa). In 1952, the Museum of Modern Art in New York organized the exhibition 'Olivetti: Design in Italy'.

and Carlo Mollino with his furniture composed of curves. A scholar, architect and engineer as well as a motor-racing and skiing enthusiast, Mollino used materials to their best effect and created arabesques in moulded plywood.

With its culture and an efficient system of production as its main assets, Italy threw itself into industrial design. It was not subject to any outside influences. It adopted a bold attitude, turning out designers such as Marcello Nizzoli and Marco Zanuso, who worked in the area of small-scale equipment such as typewriters, calculators, radios and television sets. The *Isetta,* a vehicle designed by Mario Preti for Iso (1955), the *Vespa* by Corradino d'Ascanio for Piaggio (1946) and the *Lexicon 80,* a typewriter by Marcello Nizzoli for Olivetti (1948), were based on rational design with gently curving, harmonious lines. The design movement had been launched and was embraced enthusiastically by the public. These years of optimism reached their height in 1958. With the economic boom, designers could no longer be kept outside the system: they had to become involved in merchandising.

France discovers industrial design

In 1947, the International Exhibition of Town Planning and Housing was held in Paris, organized under the auspices of the Ministry for Reconstruction. Its aim was to encourage the creation of the mass-produced furniture required for reconstruction and new housing. In spite of the exceptional projects that were presented, the exhibition was not a success.

The Salon des Arts Ménagers

Used for the promotion of mass consumerism, technological rationalization impinged on every sector of daily life. The American way of life influenced people's habits; the idea of comfort provided by efficient home management was taking hold. The first exhibition in Paris of hitherto unheard-of appliances was in 1923: washing machines, vacuum cleaners, cookers and irons were displayed on the Champ-de-Mars by Jules-Louis Breton. The Salon des Arts Ménagers (Ideal Home Exhibition) was established in 1926. The first Salon of the post-war period, in 1948, was a huge success.

The French people wanted to dream. Thanks to the building of large residential estates, most of them could be decently housed. The first area to see innovations was the kitchen, with the idea of a fully fitted 'rational kitchen', a laboratory for the modern woman. The Salon des Arts Ménagers was an extraordinary breeding ground and showcase for techniques and ideas. Each year it focused on a specific theme: chairs (1952), rattan chairs (1954), plastic furniture (1955), dining tables (1956) and bookcases and wall shelving (1957). The Salon also organized competi-

Jean Prouvé

Jean Prouvé, chair made of folded sheet metal with a seat of moulded wood, after the 1935 model. Made at the Ateliers de Maxéville in 1951.

In the period of post-war reconstruction, the ironworker Jean Prouvé (1901–84) came into his own. The introduction of electric welding and stainless steel in 1925 had made it possible to produce objects that matched the architecture of the day. From 1928, it became Prouvé's ambition to manufacture mass-produced furniture. He did not want tubular steel; he was inspired by steel plate, whether swaged, folded, ribbed or welded. In 1930, he met Robert Mallet-Stevens and Le Corbusier, both members of the UAM (Union des Artistes Modernes). He made furniture for the university halls of residence in Nancy, using a combination of wood and metal (1933), then created furniture for the Paris Electricity Company (1935). He collaborated with Eugène Beaudoin and Marcel Lods in designing the Maison du Peuple at Clichy (1938), a technical and architectural landmark. During the war he was a member of the resistance, becoming mayor of Nancy when France was liberated. In 1944 he founded the Ateliers de Maxéville, a company making furniture and prefabricated elements for building. The workshops experienced a period of intense activity lasting until 1953: 150 people worked there as a team. The Ministry for Town Planning commissioned 800 prefabricated houses in 1945. Prouvé designed them, but the inadequate industrial stock of steel held the project up and he turned to aluminium. In 1949, the Ministry commissioned 25 houses made of aluminium from him. With Charlotte Perriand he created the library for the Maison de la Tunisie at the Cité Universitaire (accommodation for Tunisian and French students) in Paris in 1952. The colour scheme for this was chosen by Sonia Delaunay. He collaborated with Le Corbusier on the scheme to produce a prefabricated flat (1946) and participated with him in the design of the furniture and constructional elements for the Cité Radieuse project (1949) in Marseilles. The Pechiney company took over the running of the Ateliers de Maxéville in 1947, and closed the factory in 1954.

tions and displayed plans for projects. In 1952, the exhibition 'Design for Use' was organized by the cultural services of the US embassy and showed items manufactured by Herman Miller and Knoll. In 1955, a competition was held for French mass-produced furniture: 'Overall review of wooden, mass-produced furniture that is well presented, of adequate quality and reasonably priced'. The group ARP (Guariche, Mortier and Motte), Louis Sognot and René-Jean Caillette did well. In 1956, the *All-Plastic House*, a concrete application of the intrinsic qualities of plastics, was exhibited. The consultant architect was René Coulon, the architectural design was by Lionel Schein, and the interior and the decoration were by Alain Richard. In 1957, the *Kitchen of Tomorrow* was presented. It was made in the USA by General Motors. The prototype was created by Frigidaire, using scientific advances, borrowing from electronics and harnessing ultraviolet rays and induced cur-

Left. Salon des Arts Ménagers, held under the vaulted roof of the Grand Palais, Paris, 1952.
Right. Salon des Arts Ménagers, 'modern kitchen' stand, Eskal, Paris, 1954.

rents. In 1958, the 'Sahara House' designed by Jean Prouvé, Charlotte Perriand, Guy Lagneau and Piotr Kowalski provided a unit for relaxation and a unit for family life. In 1957, Charlotte Perriand presented the 'Japanese House', containing items selected by the Takashimaya department stores in collaboration with Junzo Sakakura and Sori Yanagi: it was an eye-opening demonstration of mass construction using standardized elements.

Formes Utiles

Created in 1949, the Formes Utiles movement sprang from the UAM, whose objective was to research into forms and perfect them. The opening exhibition of the UAM, at the Musée des Arts Décoratifs in the Pavillon de Marsan in Paris, was entitled 'Formes Utiles, Objets de Notre Temps'. The exhibition was based on a proposal by Francis Jourdain: as early as 1929 he had put forward the idea of displaying objects in categories as though in a department store. Concrete shape was thus given to debates about the role of the utilitarian object, its aesthetics and its function in industrial production. In the movement's 1955 manifesto, Hermant offered a formula: 'We call the form of an ordinary object or a building useful when there is an exact correspondence between its effectiveness of use, the economy of the materials used and its appeal to the feelings and the mind – or when its appearance reveals an exact balance between its function, its structure and its meaning.'

Industrial aesthetics

In 1953, the Institut d'Esthétique Industrielle (founded by Jacques Viénot) published the journal *Design Industriel* and created the 'Beauté France' label. Many French industrial designers had trained as technicians or engineers; spurred on by Viénot, and with the country opening up to the rest of the world, France finally embraced industrial design wholeheartedly. The need to sell goods led to developments in poster art and printed advertising material. Advertising attracted customers, all consuming what the advertiser manufactured; as such, it proved itself to be the corollary of mass production.

A car that everyone can afford

Citroën *2CV*, 1948.

The Citroën *2CV* was exhibited at the 1948 Car Show in France and proved enormously popular. The idea of this car dated back to 1935, when the brief was drawn up by Pierre-Jules Boulanger: 'four wheels under an umbrella, an economical, safe vehicle, capable of transporting four people and 50kg of luggage in maximum comfort'. He entrusted the project to the designer Flaminio Bertoni and the engineer André Lefèbvre; they came up with the Citroën *TPV* (1936). In 1938, some 20 models were being tested. The car was put into production in 1949 with an output of four vehicles a day, rising in 1950 to 400 a day.

Ferdinand Porsche (1875–1951) dreamt of building 'the people's car'. With the backing of the Third Reich, his dream became a reality: the *Volkswagen* (People's Car), designed by Erwin Komenda, was ready in August 1938. But World War II intervened, and the *Volkswagen* (or 'Beetle', as it became known) did not go into production until 1946. 1,000 cars were manufactured in 1946, 100,000 in 1950 and 1,000,000 in 1959.

In 1942, there were about 26 million cars on the road in the USA. However, the car industry halted its research as soon as the country entered the war. Innovation did not return until

Volkswagen type 60, 1938.

Morris Minor, 1948.

1948, with a distinctive alteration to the shape of the bodywork. Cadillac introduced the biggest changes in style: Franklin Q Hershey designed the *Series 61* and *Series 62* town coupé. Taking his inspiration from Lockheed's *P-38* double-fuselage plane designed by Kelly Johnson in 1939, he gave the rear wings a fishtail shape and added a gleaming, curved radiator grille. It was an immediate hit, with the undulating line accentuated by the fins giving the impression that the car was bigger than it was.

In Britain, the great innovator was Sir Alexander Arnold Constantine Issigonis, who designed the *Morris Minor* and the *Mini*, which embodied his philosophy: 'the maximum amount of interior space for minimum external size'. His prototype the *Mosquito* had a monocoque body, 14-inch wheels instead of 17-inch ones and a low engine that brought down the centre of gravity and thus ensured better stability. In order to avoid road tax based on large engine size, Morris reduced the engine capacity to 800 cc. Issigonis also reduced the proportions of the *Mosquito* and came up with the *Morris Minor*, which went into production in 1948. Simple and cheap to run, it was the most advanced of the European small cars at that time.

Cadillac *series 62*, 1947.

The Pop years (1958–72)

- The 'youthquake'
- A new technology: plastic
- Pop Art, strip cartoons and advertising
- The acid years (1968–72)
- Sensitive environment, interior landscape
- The design object

In the late 1950s and early 1960s, a flamboyant movement began to emerge in Britain and the USA which abandoned the rational orthodoxy of 'good design'. Drawing on the colourful, cartoon-inspired imagery of Roy Lichtenstein and Andy Warhol, the Pop decade made use of vivid colours and full forms. Designers assimilated the graphic world of poster art and advertising into their work. A new energy was harnessed to serve mass consumerism, as a real revolution occurred in popular taste during the 1960s. Its success was considerable, and market pressures became inescapable.

The 'youthquake'

The 1960s marked a turning-point: it was a period when cultural and social attitudes, values, standards and opinions were transformed. Less than a decade after the end of rationing, major changes took place. Personal comfort increased, and consumption increased with it. In 1963–73, 15- to 19-year-olds represented 8% of the population for the first time in a century; the proportion stabilized at 7% towards the end of the 1970s. The media were focused on the young, their fashions and their idols. The teenage generation left its mark everywhere. Teenagers expressed their discontent and challenged the cultural and social values of their parents. They showed a desire for change, rejecting functionalism and calling for design that used colour and expressive forms.

Pop design

The young chose anti-functionalist design as an ideal expression of their values. From the start this was conveyed by a profusion of objects and an anti-conventional lifestyle. The new attitude they favoured expressed itself in music, attitude and clothes. Many designers saw this as the source of a possible renaissance of design. They exploited the ideas that arose from this movement, and businesses got involved. The baby-boom generation was targeted by manufacturers. It was not a matter of defining a style, as had previously been the case, but a fluctuating, ephemeral, free, variable movement, without criteria or standards. The most decisive changes took place in Britain and the USA. In Britain, 1962 marked the beginning of a period of prosperity after the economic recession of 1960–61.

The new generation of designers used what would later be called Pop aesthetics in the design of clothing, furniture and sometimes even architecture. Their visual sources were directly derived from Pop Art, which borrowed many elements from popular culture, and later from 'Op Art', a movement in painting which had emerged on both sides of the Atlantic. This used visual references inspired particularly by the aesthetics of space research, which had caught the popular imagination of the time.

Previous page:
S chair in polyurethane, Verner Panton, manufactured by Vitra, 1967–8. Paris, CCI-Georges Pompidou Centre/Kandinsky Library.

Towards the mid-1960s, an explosion of images, icons and symbols announced a new trend in design, which drew on the arbitrary and the ephemeral. Moreover, with the introduction of the mini-skirt and jeans, styles had to change – for one thing, people no longer sat down in the same way. The traditional relationship between form and function was replaced by a more extrovert relationship between form and expression.

Throwaway and ephemeral

The age of gadgets got off to a start with ingenious, very cheap objects: paper bags with pictures on them, wrapping paper, boxes and mobiles. Some designers came up with throwaway dresses made of paper with pictures on it. Products became obsolete very quickly. Publicity and marketing campaigns encouraged this process: goods were made to be seductive and create a demand, not necessarily to respond to a need. The throwaway principle even found its way into the furniture sector – hitherto regarded as a producer of durable products – with the use of cardboard and inflatable PVC.

Furniture was seen as flexible rather than as a monumental sculptural element. Young people needed furniture that expressed their way of life, and cardboard furniture was a good reflection of the shifting mood of youth and its taste for the ephemeral. The use of this type of furniture started in the 1960s. The British designer Peter Murdoch came up with a chair that became a real icon of Pop furniture: the *Polka-Dot* chair (1963), an armchair for children made from paper. It was the first cardboard item of furniture to be put on the market. It was made from three sorts of paper bonded together in layers to create pasteboard and coated with a seal that made it washable and relatively rigid. It had a lifespan of about three to six months. Sold flat, it was decorated with coloured dots and brightened up the shelves of department stores and supermarkets. It was space-saving, tough and cheap. The polka-dot pattern was applied in a single operation when the cardboard was laminated. New colours or patterns could be introduced without altering the system. The chair was assembled at home by following the folding instructions. In 1966, it attracted international attention, but the designer could not find a manufacturer in Britain. Eventually the International Paper Corporation in the USA put it into production. This icon of Pop furniture – remarkable for its low production cost and its easy availability – was an ideal object for mass consumerism.
Commissioned by the 1966 Ideal Home Exhibition, the British designer Bernard Holdaway produced a range of sturdy, washable furniture made of reinforced cardboard. He exhibited a suite of furniture for the family home, using tubes and flat sheets of compressed cardboard.

Holdaway's aim was to produce an attractive design at a minimum price. Bright colours – red, blue, yellow, brown and purple – gave the furniture a Pop look, with cushions in complementary colours. Known as

Polka-Dot chair for children, Peter Murdoch, 1963. Throwaway folding armchair made of cardboard, decorated with a pattern of coloured dots. It was produced cheaply in large quantities and sold flat. Manufactured by the International Paper Corporation, USA.

the *Tomotom* range, it was featured in design magazines and produced by a small company called Hull Traders. British designers and consumers were all for experimentation. In January 1967, the Design Centre in London held an exhibition of prototype furniture. Many of the items displayed, made by students or young designers, used plastic or cardboard. In France, Jean-Louis Avril (b.1935) designed a suite of furniture made from painted millboard: seats, pouffes, storage units and a pole lamp (1967). He then moved on to dining-room and bedroom furniture. His pieces were marketed in two versions, varnished natural millboard or painted millboard. Claude Courtecuisse also explored this throwaway material, using folded corrugated cardboard to make furniture decorated with painted patterns (1967).

A new technology: plastic

The furniture industry responded to the need for large-scale production in order to satisfy the tastes of the baby-boom generation and provide them with furniture that was inexpensive, mobile, light and colourful. Given the cheapness of petroleum in the early 1960s, it is not surprising that the industry turned to plastics. ABS (a rigid plastic), polyethylene and other thermoplastics were the materials of the decade. These synthetic materials had a many good qualities: lightness, toughness, the possibility of being produced in colour, and a shiny surface on both sides. Although the raw material itself was cheap, the making of moulds was expensive, and so items were produced in large runs to make investment worthwhile. The products were sold at a reasonable price. They had to sell quickly before, as was bound to happen, they were replaced by other, more fashionable products. Industry was forced to focus on the notions of marketing, consumer products and changing fashion.

Italy leads the way

Italian manufacturers were ready to respond to the demands of mass culture. Many new companies were set up, including C&B, Kartell, Poltronova, Artemide and Zanotta. The most important contribution made by Italian industry was perfecting the technique for injection-moulded plastic. In the field of plastics, the Montecatini Research Institute produced an experimental bathroom designed by Alberto Rosselli in 1957, while Roberto Menghi designed a set of plastic containers made of polyethylene, a material with qualities well suited to the design of objects. The company Kartell, in collaboration with the designer Gino Colombini, created a series of household objects, including dustpans (1958) and plastic sieves (1962).

From 1960 onwards, Marco Zanuso produced a large number of high-quality objects without being unduly concerned with fashion. His work was very much geared towards technological innovation, which is how he came to make the first plastic stacking child's chair, model *4999*, with Richard Sapper in 1961, after four years of research into the new plastics technology. It was manufactured by Kartell, and was awarded the Compasso d'Oro in 1964.

At the Seventh Furniture Exhibition in Milan in 1967, plastics had proved themselves, if the comments made in *Home Furnishings Daily* (New York, 5 October 1967) are anything to go by: 'Plastics have inspired the Italians and given them new creative opportunities as regards weight, volume, surface, form and flexibility – even for the most massive of forms. Indeed, plastics technology inspires them so much that they are using their artistic virtuosity to produce furniture that has never been made before.' These comments refer to Joe Colombo's *4867* chair for Kartell, which was directly derived from Zanuso's and Sapper's *4999* chair.

Kartell's stand at the Furniture Exhibition was all green and white, with storage modules made from ABS plastic and designed by Anna Castelli Ferrieri. These could be stacked in various ways: one element made a stool, two elements made a coffee table, and four elements made a set of shelves. The *Gaia* stacking armchair by Carlo Bartoli, made from polyester and fibreglass, was also on show.

Vico Magistretti made rational use of plastic with the *Selene* chair (1969), manufactured by Artemide. Joe Colombo produced several plastic chairs at the beginning of the decade, including the *Elda* armchair (1963), marketed by Comfort, and the *4867* stacking chair (1965). Achille and Pier Giacomo Castiglioni designed low table-cum-stools made from polyester (1965).

By 1965, Italian manufacturers expected that consumption would double in the years ahead. This explains the very rapid development of Italian design and manufacturing output and the international recognition accorded them. Exhibitions and magazines recorded the extraordinary activity – especially where design for the home was concerned – of Magistretti, Colombo, Gae Aulenti, Tobia and Afra Scarpa, Giotto Stoppino, Cini Boeri, Gianfranco Frattini, Sergio Asti and Massimo Vignelli. These designers explored the qualities of plastic materials inventively and elegantly. They ushered in a new culture of design for the home that remains the greatest achievement of Italian design. Each of them introduced a new approach, using various means of poetic expression.

Plastic in other countries

In 1963, the British designer Robin Day exploited the advantages of polypropylene in his design for the *Mk2* stacking chair, manufactured by Hille. Combining elegance and durability, it was used throughout the world by community groups and institutions. The seat and the back were injection-moulded in one single, impact-resistant piece, made of material that would not deteriorate. Impact resistance tests were carried out in the laboratory. The basic model was stackable.

France developed the use of plastic in a very craft-based way. Towards the end of the 1960s, several designers were showing an interest in it. Christian Germanaz designed the *Half and Half* seat (1964), manufactured by Airborne in 1968. This consisted of two identical plastic shapes clamped together to form a bench. In 1969, Marc Held created a collection of furniture made of polyester and fibreglass for the architect Georges Candilis. In 1970, this furniture was sold through the Prisunic supermarket chain. Marc Berthier designed the *Ozoo* desk and chair in 1965, when he was director of the design office at Galeries Lafayette; his collection, made of polyester reinforced with fibreglass, was manufactured by Roche-Bobois in 1967.

In the USA, Wendell Castle designed biomorphic plastic furniture. In the Scandinavian countries, the Finnish designer Eero Aarnio embarked on research into plastic furniture. The Asko company manufactured both of his models: the *Globe* chair (1963), which derives from the world of

Verner Panton

S chair, made of polyurethane, Verner Panton, 1967–8, manufactured by Vitra. This model takes up the idea of Gerrit T Rietveld's *Zig-Zag* chair (1934), as reinterpreted by Panton in 1955 and manufactured by Herman Miller. Paris, CCI-Georges Pompidou Centre/Kandinsky Library.

Verner Panton (1926–98) was a Danish architect who studied architecture at the Royal Academy in Copenhagen. Between 1947 and 1951, he was influenced by Arne Jacobsen and Poul Henningsen, for whom he worked. He then embarked on a career as a designer, preferring the challenge of new technologies to the craft traditions. His first 'manifesto' was the *Cone* chair (1958). He made use of polyurethane in 1960 with his S-shaped chair, based on a design with a fluid form. The development of the prototype took several years, before it was manufactured by Vitra in 1967–8 with the technical assistance of Herman Miller. He created other chairs during the 1960s, but in particular he concentrated on developing his ideas on interior design. He carried out the successful complete renovation of the Astoria Hotel in Trondheim (1960), taking great care to select the right forms, patterns and colours for everything, from floor to ceiling. Panton was primarily interested in new materials and the new techniques that went with them. He enjoyed experimenting, and even if some of his projects never went into production, they added to the accumulated store of design knowledge. His *Phantasy Landscape*, an environment commissioned by Bayer (1970), was a sensuous, intimate universe made of foam, where light and colour were intimately connected and where it was possible to live at floor level, lying down or seated. The *Living Tower* (1968) was already indicative of this change of direction. He designed not only the furniture, but also the carpets, curtains, lighting and mural decorations, all of them with Op Art motifs. Panton always sought to create an environment where the relationships between the colours, forms and light were glaringly apparent. He set up his agency in Switzerland and won numerous awards, including the International Design Award in 1963. He exhibited at the Eurodomus in Turin in 1968.

Globe or *Ball* chair, fibreglass, Eero Aarnio, 1966. Manufactured by Asko, Finland. This chair represents 'a room within a room', a snug, sheltered space. Vitra Design Museum.

science fiction, and the nature-inspired *Pastilli* armchair (1967).

Both of these designs were produced in brightly coloured fibreglass, and became real icons of Pop furniture. The English fashion designer Mary Quant bought a Globe chair for her London boutique in 1967. It also featured in films and television series.

Pop Art, strip cartoons and advertising

Pop Art was based on an observation of street life. The American artists Andy Warhol, Claes Oldenburg and Roy Lichtenstein and the French artist Martial Raysse took its colours, characteristics, objects and neon signs and put them centre-stage. A second generation of Pop artists emerged in England around 1961. Most of them had studied at the Royal College of Art, and they were reunited at the 'Young Contemporaries' exhibition: Peter Phillips, Derek Boshier, David Hockney and Allen Jones. The figurative imagery of these artists had its source in racing cars, motorbikes, spaceships, sex symbols, pinball machines and neon signs. As far as youth culture was concerned, England – with Mary Quant and the fashion for mini-skirts – reigned supreme.

Pop icons, Pop places, Pop personalities

Pop icons were adopted in England by Binder, Vaughan and Edwards, who customized furniture – drawers and chairs – with psychedelic patterns. Their clients included Lord Snowdon, David Bailey, Princess Margaret, Henry Moore and John Lennon (they decorated a piano for Lennon in 1966). Using flamboyant colours, they customized a Buick convertible and an AC Cobra car, which were then displayed at the Robert Frazer Gallery in London. John Vaughan started to paint murals for Lord John in Carnaby Street (1967), but his work did not meet with the approval of the authorities. He occasionally fell foul of the law: for example, he was not allowed to paint on walls, which was viewed as graffiti. Jon Bannenberg designed a Pop Art interior for Mary Quant and her husband in 1965. The Mr Freedom shop was another legendary place; the designer John Weallans left his mark here in 1968–9, decorating it with neon lights, pictures of Mickey Mouse and murals in the spirit of Roy Lichtenstein. It was not only a shop, but also a meeting-place with a café: an anarchic place where Elton John rubbed shoulders with Peter Sellers.

In the USA, Andy Warhol embodied the Pop spirit. He was based at the Factory, a famous New York address which was simultaneously a studio, a film set and a venue for wild parties. It was a highly influential model of decadent living and an international meeting-place. Warhol found himself in the vanguard of the movement to rehabilitate industrial premises: a 'loft' spirit was prevalent at his studio long before lofts became fashionable. Billy Name, one of the Factory personalities, provided Warhol with inspiration for the decoration of the Factory. He covered every surface, from floor to ceiling, with silver paper. This concept of an improvised, anti-aesthetic space was the complete antithesis of the ultra-sophisticated New York interiors in fashion at the time.

Do-it-yourself

Interior decoration was free of dogma and doctrine. Its watchword was 'do-it-yourself'. By using techniques such as transformation, collage, improvisation and the juxtaposition of colours, people could use decorative elements again and again, changing the look of a room as often as they liked.

Furniture was now sold by mail order, which meant that it was no different from other consumer goods. It was no longer regarded as an exceptional investment. From now on, furnishing the home was as ordinary as buying clothes. Terence Conran was the first to use this sales method in Britain.

In France, Maïmé Arnodin and Denise Fayolle used the same principle for the Prisunic supermarket chain in 1968. They produced a catalogue of furniture designed both by their own design agency and by great designers: Joe Colombo, Gae Aulenti, Marc Held, Olivier Mourgue

Variations on the *Culbuto* armchair, made of polyester and fibreglass, in Marc Held's workshop in Paris in 1972. The stacking moulds of the various elements of furniture are hanging on the wall.

and Marc Berthier. These products were new and popular. They were geared towards to the needs of young people and could be afforded by those on low incomes.

Practical, flexible, stackable furniture

Furniture and design underwent a great many changes in this period. Government offices, hospitals and hotels had an urgent need for furniture that was easily transportable in quantity, easy to handle and cheap. This notion of flexibility was inherent in knockdown furniture, and many designers would respond to the requirements of the time.

In Britain, the company Race produced Nicholas Frewing's *Flexible Chair*, which can be assembled in just a few minutes, and Max Clendinning's *Maxima* furniture (1965), consisting of 25 standard elements that can be put together in 300 different ways. The system offered great flexibility as well as a trendy image.

The modular system – based on the principle of standard elements that can be assembled in various combinations – was developed at the Furniture Exhibition in Italy, with innovations appearing every year: in 1966, Eugenio Gerli's *Domino* shelving system, manufactured by Tecno,

Habitat

Shop window of the first Habitat shop at 77 Fulham Road in London. Customers were offered a mixture of styles, including traditional English and Finnish design.

In May 1964, the designer and businessman Terence Conran sparked a revolution in designer goods retailing when he opened the first Habitat shop at 77 Fulham Road in London. His involvement in design went back to 1955 with the founding of the Conran Design Group. With Habitat he was selling a concept of a 'total lifestyle'. Responding to young people's aspirations to furnish their homes in their own style, he offered them a varied selection of goods so that they could plan and decorate their homes to suit their own taste. He came up with well-designed furniture that was both trendy and traditional. Habitat sold its own designs as well as reproductions of furniture classics such as Marcel Breuer's *Cesca* armchair or Le Corbusier's chairs. It also sold lighting, toys, kitchen utensils, wallpaper, curtain fabrics and china.

The Habitat style appealed to the young. It offered a mixed selection of products, combining Scandinavian-style good taste with elements taken from traditional English style, and blending antique and modern elements. With its white painted walls and red tiles, the shop gave customers ideas, encouraging them to make design an integral part of their own homes. The goods were sold on a self-service basis, simply piled up on the floor or arranged together on shelves. Customers could browse in pleasant surroundings. A second Habitat shop opened in Tottenham Court Road in 1966, and a further five new shops opened in the late 1960s. Habitat was the first company in Britain to use mail-order selling. The catalogue had a modern look and proved very successful.

and Roberto Pamio's flexible bed *Cleo*, manufactured by Esse; in 1967, C&B exhibited Marco Zanuso's *Lombrico*, a sofa consisting of a series of elements – foam seating units that could be assembled, with just a screwdriver, to form a sofa of any length.

Manufactured by Poltronova, Ettore Sottsass's *Kubirolo* units – 45 x 45 x 45cm beige or white plastic units with red, yellow or black handles – were designed to be put together in various combinations. Similarly, C&B produced a system of panels that could be joined together at right angles. The company Design Center brought out *Cub 8* by Angelo Mangiarotti, which came with PVC linking pieces to assemble the panels. Another modular system, Tito Agnoli's *Programma C*, was manufactured by Citterio. In 1969, Joe Colombo used the principle of sets of cushions of different sizes for his *Additional System* seating, manufactured by Sormani.

In the USA, Herman Miller continued its research into office furniture with Action Office 1 (1964) by George Nelson and Robert Propst. The system – consisting of a vertical storage unit, a switchboard and a desk – was totally interchangeable. It marked a major advance in this type of furniture. Robert Propst took the project further by creating *Co-Struc* (Coherent Structures Hospital System, 1969), a system of containers, frames, sliding doors and panels that could be combined in an endless number of ways.

Pop did not bring all the changes its initiators had hoped for, but it did make a difference to people's lifestyles and increase the design choices available to the young. The real revolution took place in Italy and London. Between 1963 and 1967, Carnaby Street came to epitomize 'Swinging London'. The song 'My Generation' by The Who (1965) supplied one of the slogans of the young generation: 'Hope I'll die before I get old'.

The acid years (1968–72)

The year 1968 marked a turning point: the Soviet Union deployed its troops in Czechoslovakia, Martin Luther King and Robert Kennedy were assassinated, Richard Nixon became President of the USA, young people woke up to politics, and in May radical political action was taken by students in France and elsewhere. The words anarchy, radicalism and agit-prop were on everyone's lips. It was the end of the Pop era, the end of a period of consensus.

Psychedelic culture and the conquest of space

Young people developed a new attitude. They were bored with opulence and passionately interested in politics. In the USA they protested against the Vietnam War. Minorities rose up and fought against social injustice – or escaped from it: the drug culture and hippies appeared on the scene

Set of armchairs from the *Djinn* series, Olivier Mourgue, 1964, manufactured by Airborne. This furniture featured in Stanley Kubrick's science-fiction film *2001: A Space Odyssey* (1968).

towards the end of the decade. The use of LSD, or 'acid', spread throughout Britain and the USA. The 'high priest' Timothy Leary urged people to take a trip towards spiritual ecstasy. Those who followed his teachings – hippies – had their own lifestyle, and hallucinogenic drugs were their path to nirvana. This naive optimism was represented by the British singer Donovan.

A psychedelic environment of undulating forms and vivid colours spread into every domain: clothing, posters, furniture, record sleeves and murals. But the psychedelic style was merely transitional.

In 1969, Neil Armstrong walked on the Moon: 'one giant leap for mankind'. The end of the 1960s was characterized by imagery associated with the conquest of space, which influenced design, fashion, furniture and objects. Courrèges produced fashions using PVC, while Rabanne used metal. Hans Hollein's boutiques were decorated in stainless steel and aluminium. Mary Quant lived in a world of dazzling white. Compact, efficient, all-encompassing designs were being developed both for interiors and exteriors. Some designers were in love with high technology, which they explored in integrated micro-environments that made use of this technology and that were inspired by the visual references of the space race. These micro-environments, looking like space capsules, were often made of plastic and brought together all the equipment needed for everyday living: kitchen equipment, radio, cassette player, television, video, and so on.

The media latched onto this design vocabulary. In Roger Vadim's film *Barbarella* (1967), Jane Fonda is living in a science-fiction world. She has a magnificent inflatable, transparent bed, and walks around in a capsule with a soft floor and inflatable walls. The director Stanley

Visiona 2, Verner Panton, 1970. An environment commissioned by the company Bayer for the International Furniture Exhibition in Cologne.

Kubrick created a prophetic vision in *2001: A Space Odyssey* (1968), one of the artistic highlights of the year. The film was first shown in London at the Casino-Cinerama on 10 May and was hugely successful. It ran in Manhattan for over a year. John Lennon was one of its admirers. The space station in the film is fitted out with Olivier Mourgue's *Djinn* series chairs (manufactured by Airborne, 1964–5).

Design and anti-design

The exhibition 'Italy, the New Domestic Landscape' (1972) at the Museum of Modern Art in New York showed micro-environments specially developed for the exhibition. In preparation for this, in-depth research into the subject was undertaken, confirming the trends touched upon throughout the decade by the work of designers such as Joe Colombo. The brief was to consider the recent history of design and prototypes, and try to find – by studying the past – harmonious solutions for the future. The schemes submitted by designers and manufacturers in 1972 displayed two opposing tendencies. The first saw design as a solution to the problems encountered in the natural and socio-cultural environments. The second was an anti-design attitude which contemplated the overthrow of the structures of society. The competitive exhibition was open to Italian designers under 35 years of age. They were asked to devise a domestic environment dictated by new consumer habits and changes in behaviour. The domestic environments shown were by Gae Aulenti, Ettore Sottsass and Joe Colombo, and the mobile environments by Mario Bellini (with his *Kar-a-Sutra*), Alberto Rosselli, and Marco Zanuso working with Richard Sapper.

Japanese designers, motivated by the shortage of accommodation,

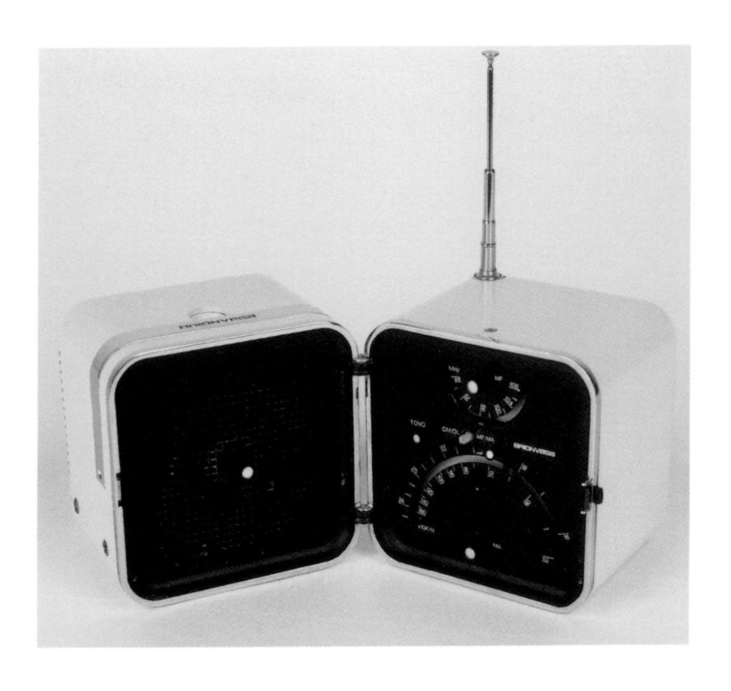

TS 502 radio, Marco Zanuso and Richard Sapper, 1965. Manufactured by Brionvega.
The technical efficiency of this cube-shaped radio is shown in the way it opens up into two halves, with the loudspeaker on one side and the controls on the other.

Joe C Colombo

Living Unit, Joe C Colombo, 1969. This interior was commissioned from the designer by the company Bayer for the 'Visiona' exhibition.

Born in 1930, the Italian Joe C Colombo studied architecture, painting and sculpture. In 1951, he joined the Art Informel movement, exhibiting with Lucio Fontana, Bruno Munari, Enrico Baj and Roberto Matta. In 1952–3, he designed his first building, which was put up in Milan. After taking over his father's car-selling business, in 1962 he returned to his work as an architect, taking a special interest in interior design. He used Plexiglas for his *Acrilica* lamp (1962). His lights, armchairs and furniture were selected for the 13th Milan Triennale, where he was awarded three medals: for his *Projector*, *Mini-Kitchen* and *Containers*. Colombo was enthusiastic about modular living systems, such as the *Combi-Centre* (1963). His furniture – for instance, the *Universal* chair made of ABS (1965) – was acquired by major museums. His *4867* plastic chair, manufactured by Kartell (1967), was his first experiment with the use of the new plastics technology. The *Uno Box* (1968), manufactured by La Linea, condenses a fully fitted bedroom for one person into a single element: bed, storage units, desk, lamp and a ladder to climb up to the bed. He studied ergonomics, sociology and marketing, and calculated the dynamics of the contemporary home. His programmable *System for Living* was exhibited in 1968. Made by GM Arredimenti, it was intended for distribution by the Rinascente department stores in Milan. The *Tube-Chair* (1970), manufactured by Flexform, is a set of four concentric tubes with different sections that can be fitted one inside another for ease of transportation. By using U-shaped clips, armchairs or sofas can be created from the tubes. In 1969, the year men walked on the Moon, Colombo presented his Visiona 1 programme, *Roto-Living* and the *Cabriolet-Bed* – exceptionally high-tech, open-ended projects. In 1971, the year of his death, he exhibited the *Total Furnishing Unit* at the Museum of Modern Art in New York. This, the final result of his thinking about compact living units, grouped the functions of the house according to specialized sectors.

carried out research along similar lines. The architect Kisho Kurokawa presented a tower of fully fitted living-pod apartments at the Osaka World Fair of 1970. This idea of the compression of space and functional efficiency derived from astronautics, and was influenced in particular by the design of the living quarters of rockets.

Sensitive environment, interior landscape

The British group Archigram instigated a protest movement against the excesses of technology. In their *Plug 'n' Clip* urban project, they tackled the problem of how to organize an interior. They designed a three-dimensional space comprising only a floor, ceiling and walls, with integral furniture made of inflatable plastic: a comfortable space with no rough edges. Inflatable furniture was subjected to experimental research. Using synthetic materials associated with space research, the furniture took on an ephemeral character. It could be inflated if friends arrived, or folded away to free up more space. Inflatables fitted in well with the idea of a utopian architecture.

Inflatables

Verner Panton was the first person to come up with an inflatable pouffe in 1962. In 1964, prototypes were developed by Cedric Price, but it was the Italian designers Paolo Lomazzi, Carla Scolari, Donato D'Urbino and Jonathan De Pas who designed the *Blow* armchair (1967), the first inflatable armchair that was popular and affordable thanks to a new plastics technology using radio-frequency welding to seal the seams. The armchair was bought flat and inflated at home. Its contours were reminiscent of certain modernist armchairs of the 1930s, especially those of Eileen Gray. It became a real Sixties icon.

In France, the 1967 Salon des Artistes Décorateurs and then the ARC (Atelier de Recherche Contemporaine) at the Musée d'Art Moderne in Paris exhibited inflatable structures. The French group Aérolande, in collaboration with Aubert, Jungman and Stinco, developed a proper strategy for publicizing inflatables and were especially active at the ARC's exhibition in 1968. They demonstrated the wide variety of products that could be made by using the inflatable technique: these included buildings, boats, barriers, temporary structures, mattresses, furniture and games. Inflatables offered a new, 21st-century design. At Expo 67 in Montreal in 1967 and at the Osaka World Fair in 1970, many inflatable objects were shown. Quasar Khanh produced a range of inflatable furniture, including a sofa and an armchair, both coloured and transparent, with metal rings to hold the elements in position. The company Ultra-Lite manufactured Khanh's models in Britain. In 1968, it was turning out 300 items a week. There were many articles in the press singing the praises of inflatables, in particular their low cost.

Non-furniture

As traditional furniture could not fit into this unconventional environment, designers – inspired by articles such as 'Chairs as Art' by Reyner Banham in *New Society* (20 April 1967) – invented furniture with forms that were expressive rather than functional.

Models were required that could be used more flexibly, and that were indicative of changing attitudes.

Informal horizontal surfaces made their appearance. For Max Clendinning, the new style of furniture was an interchangeable combination of cushions. Seats for lying on rather than sitting on were designed in Italy, France, Scandinavia and Britain. Roger Dean's *Sea Urchin* was shown in the 'Prototype Furniture' exhibition at the Design Centre in London in 1967. Hille considered putting this into production, but technical problems made mass production impossible. In France, Pierre Paulin came up with ideas for comfortable seating: the *Ribbon Chair* (1966) and the *Tongue Chair* (1967), which had a metal structure upholstered in foam. The Italians were very bold and original in their approach, not working from any preconceived ideas. Their designs were sometimes shocking in their freedom and anti-conformism. '[There is] a

Blow inflatable armchair, PVC, Jonathan De Pas, Paolo Lomazzi and Donato D'Urbino, 1967. Manufactured by Zanotta, Italy.

Ribbon Chair 582, Pierre Paulin, 1965. Manufactured by Artifort, Netherlands. The base is made of steel and the tubular structure is covered with foam and upholstered in a coloured stretch fabric. Vitra Design Museum.

need to place furniture on the same level as society and speed up its development as we head towards 1970, or indeed 2000,' Jean Daniel told *L'Officiel de l'Ameublement* (November 1967).

The painter Roberto Matta designed *Malitte* seating (1966) for the manufacturer Gavina – a sculpture of foam seats that could be arranged in different combinations. When they were not being used, they could be stacked on top of one another, forming a vertical ornamental sculpture. At the Milan Furniture Exhibition, Archizoom showed the *Superonda* manufactured by Poltronova, which was a modular divan along the same lines. It consisted of a block of polyurethane foam covered with oilcloth cut in wave shapes, suggestive of freedom. The *Superonda* had one novel feature: its two sections could also be used as a sculpture, two sofas, an armchair or a table The most surprising thing about these new models was the notion of furniture without any structural frame – indicative in itself of a rebellion against the established order.

'Up! Up! Up!'

An article by Henry Owen in *Cabinet Maker and Retail Furnisher* (London, October 1969) proclaimed: 'Up! Up! Up!!! in Milan'. It referred to the novel items exhibited by C&B at the Milan Furniture Exhibition in 1969: the *Up* series by Gaetano Pesce. This series made a strong conceptual impact, and in terms of technological innovativeness it was quite remarkable. The *Up* seat was sold in a flat vacuum pack. All that was needed was for the cardboard box to be opened, and the seat – made of honeycomb polyurethane foam upholstered with a mixed viscose, nylon and Lycra fabric – gradually sprang back into shape. Speaking in 1993, Gaetano Pesce said: 'When people try to understand a concept that is not obvious, each person interprets it in his or her own way. In the case of *Up 5*, some people interpreted the armchair as the body of a mother who always welcomes us lovingly and cradles us in her arms. Others said it was a sexual symbol. Others said many other things. I found it good that the object was open to different interpretations. My own interpretation is that it represents someone who isn't free, a prisoner walking with a ball chained to their leg. Throughout history women have always suffered as a result of men's prejudices. It's like being in prison, like walking with a weight on your leg all the time. The only way of expressing this concept was to exhibit the ottoman, or the ball attached to the female body by a chain.'

Young people were also very keen on big reshapable cushions, available from Habitat or elsewhere at reasonable prices. Alternatively, people made them themselves. In Denmark, Nanna Ditzel devised an interior structured by landings, with cushions, steps used for storage, sofas and

Up 5 or *La Donna* armchair, Gaetano Pesce, 1969. This series, made of jersey-covered polyurethane foam, consists of seven different models, sold in vacuum-pack form. Manufactured by C&B, Italy. Vitra Design Museum.

Sacco

Sacco chair, Piero Gatti, Cesare Paolini and Franco Teodoro, 1968. Manufactured by Zanotta, Italy.

The *Sacco* chair (1968) by Piero Gatti, Cesare Paolini and Franco Teodoro, manufactured by Zanotta, introduced a revolutionary idea appropriate to the new way of life. It was a vinyl or leather bag filled with polystyrene foam beads. At the time it was completely original, but it has been much imitated since. It went back to the ancient idea of sitting on a pile of sand. Containing 12 million polystyrene beads and weighing less than 6kg, it dispensed with all the obvious elements of a chair (underframe, seat, back) and as a result was an object with no defined form. It was easily transportable and malleable: the body in it gives it its form and meaning. The *Sacco* would mould itself to the individual shape of the person using it. Its sensuous malleability, as it adapted to every changing posture of the body, was an invitation to relax. It let the sitter nestle into it, permitting an infinite number of positions for both day and night: 'the seat for 1001 nights – 1000 positions for the daytime, one for the night – marvellously comfortable' (to quote the advertising slogan used by Zanotta). It was snapped up by young people, to whose lifestyle it seemed particularly well suited. It was no accident that it first appeared in 1968, at a time of protest against constraints. The new generation lived down on the ground. The body was everything, and people had to let themselves go. This chair was the emblem of a generation of relaxed, fairly fit (you had to be in order to get up out of it), stress-free, jeans-wearing youngsters. Because the *Sacco* was at floor level, it gave the person sitting in it the feeling of being smaller than anyone seated on an ordinary chair. It is reminiscent of some soft-structure works by the American Pop artist Claes Oldenburg. The design is also a subtle, sensuous pear shape, close to the organic spirit of the 1950s.

'conversation pits'. For his part, Verner Panton dreamt up an interior with several staggered levels, allowing those present to converse (with several other people on different levels all at once) or else withdraw into solitude.

The design object

Technicians were interested in what lay beyond design, or meta-design. They did not lose sight of the fact that behind the prototypes created by designers lay structural, three-dimensional, operational codes. Those who produced fashionable designs also had to bear meta-design in mind. Industrial objects had to be made using moulds reduced to simple forms, so that in the event of any superficial variation prohibitive costs would not be incurred. Because of their three-dimensional visual impact, high-quality industrial objects had the capacity to challenge. The view of the world implied by their use had an influence on society. There was not a single type of design, but many. There was the design-led rationalisation of handicrafts in some developing countries; there was semi-industrial design, as in the Scandinavian countries still dominated by woodcraft and metalworking; there was plain, simple design for community associations; and there was the design of everyday objects, which could give new meaning to trinkets and gadgets.

The Centre de Création Industrielle (CCI)

At the prompting of Georges Pompidou and André Malraux, the Mobilier National (the government department in charge of state-owned furniture) became more design-conscious, and the Atelier de Création Contemporaine was created in 1964. The Musée des Arts Décoratifs was at the centre of this new initiative, staging the 1962 exhibition 'L'Objet', which called upon artists to produce works expressive of a new lifestyle. The curator François Mathey followed this up in 1968 with the exhibition 'Les Assises du Siège Contemporain'. This was a landmark event, with a large section devoted to international furniture. Pierre Paulin drew inspiration from the nomads' use of rugs to create his *Tapis-Séjour*, designed for Roche-Bobois in 1965. In 1968, he came up with the *Declive* chaise longue, a kind of rug-seat that could be transformed at will into different shapes, manufactured by Mobilier International. Olivier Mourgue followed in Paulin's footsteps with a rug-cum-seat featured in the Prisunic catalogue in 1969.

The CCI (Centre de Création Industrielle) – a new department occupying six purpose-designed rooms within the Union Centrale des Arts Décoratifs – was created in 1969. It selected the best industrial products available on the French market: kitchens, heating equipment, bathroom fittings, tables, furniture, and so on. Documentation was compiled on each product, and this could be consulted by the public and industrialists. The CCI aimed to put industrialists and designers in touch with one another, and teamed up with Formes Utiles, which had always championed design. The CCI exhibitions analysed the various sectors that the department covered. The opening exhibition in 1969 – 'Qu'est-ce que le design?' – asked a general question which was answered by five of the greatest contemporary designers.

'Qu'est-ce que le design?'

The exhibition brought together five designers of different nationalities: Italian, American, German, Danish and French. They came from industrialized countries which had made substantial technological advances and introduced formal innovations in the 1960s. Design was a central concern for them – not a luxury or an affectation, but a necessity. Each of the designers came up with his own answer to the question 'What is your definition of design?' Joe C Colombo: 'Industrial design is certainly not a style; it is functional and rational. It is the complete resolution of the problems inherent in a product which has been designed in the most objective way possible and bearing in mind the use for which it is intended.' For Charles Eames design was 'a plan for arranging elements in such a way as best to accomplish a particular purpose'. Fritz Eichler stated: 'Design is a complex set of factors which is not confined to creating an aesthetic external form, for this form is simply the visible expression of a collective creative effort. Design is an integral part of our development.' Verner Panton put it like this: 'Space and form are important elements in creating an environment, and colours are even more important. But man is still the central element.' Finally, according to Roger Tallon: 'Design is not typified by a specific conceptual activity, but by the behaviour of the originator in exercising that activity. Design is a course of action that rejects any solutions that have not been thought through, whether risky or inspired. It involves looking for information and dealing methodically with every problem.'

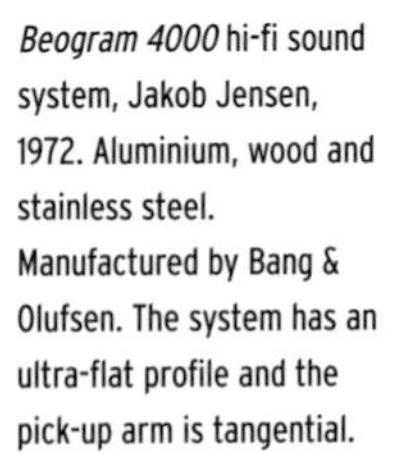

Beogram 4000 hi-fi sound system, Jakob Jensen, 1972. Aluminium, wood and stainless steel. Manufactured by Bang & Olufsen. The system has an ultra-flat profile and the pick-up arm is tangential.

Roger Tallon

P111 portable television, Roger Tallon, 1963. Designed for the company Téléavia, Italy. Paris, CCI-Georges Pompidou Centre/Kandinsky Library.

Tallon, a French designer born in 1929, studied engineering and started his career with Caterpillar France, before going on to work for Du Pont of Nemours. In 1953, he met Jacques Viénot and joined Technés, a design and drawing agency founded in 1949.

From 1954 onwards, he designed many industrial objects for various companies: a milling machine for Gambin, the *Duplex* 9.5mm cine camera for Pathé, the *Brandt-luxe B600* sewing-machine and, in 1955, the *Taon* compact 125cc motorcycle. He taught industrial design on the advanced industrial aesthetics course of the École des Arts Appliqués, the first design course in France. In 1957, he designed the Veronic 8mm cine camera for Sem and the *Gallic 14* and *Gallic 16* lathes for La Mondiale; in 1958, electrical household appliances for Peugeot; in 1959, the first television sets for Téléavia; and in 1960, transistor radios and electric razors for Thomson. He also produced furniture, designing the *Wimpy* seat in preformed plywood and cast aluminium, manufactured by Sentou. He collaborated with Yves Klein and César on the exhibition 'Antagonisme 2, l'Objet' at the Musée des Arts Décoratifs. At the ICSID (International Council of Societies of Industrial Design) conference in Paris, he projected psychedelic images on to the live-in sculpture by André Bloc at Meudon. Together with Jacques Dumond he set up the design department at the ENSAD (École Nationale Supérieure des Arts Décoratifs) in Paris. In 1963, he designed the *Japy Message* portable typewriter and, for Téléavia, the *P111* portable television. In 1965, he designed the *M400* series of chairs and stools for Galerie Lacloche. He produced his first sketch designs for the turbine-driven *TGV 001* (high speed train) for Alsthom. In 1969, he designed the *Cryptogamme* furniture for the Mobilier National, intended for use in the cafeteria at the Grand Palais. He opened his own agency in 1975, calling it Design Programmes. His career continues today, distinguished by uncompromising designs: a contemporary cast-iron spiral staircase for flats, manufactured by Galerie Lacloche in 1966; high-speed trains; Lip watches; and many others.

Éclisse table lamp with pivoting painted metal reflector, Vico Magistretti, 1967. Manufactured by Artemide, Italy.

Creative designs in Europe and Asia

In Denmark, Bang & Olufsen was one of the few companies producing radio and television equipment to care about design. Today it is the biggest Danish company at international level.

It has always kept up with the latest technological innovations. In the early 1960s, for example, it developed the tangential pick-up arm, which moves in a straight line across the record, and an amplifier with touch-operated controls. The company's design programme stressed that technology should be adapted to suit the domestic environment, and should be simple to use. Bang & Olufsen had seven 'corporate identity components', which made for a strong corporate identity. The models created towards the end of the 1960s were remarkably modern in design, thanks to Jakob Jensen (b.1926), who was in charge of styling at Bang & Olufsen. He designed the *Beosystem* (stereo loudspeakers and amplifier) and the *Beogram* (turntable) between 1969 and 1973. Switches were replaced by touch controls, and the subtle combination of dark wood veneer, satin-finish aluminium and stainless steel created a perfect balance between natural and industrial materials. In essence, it was a semi-industrial design with touches of wood and metal.

There were new developments in electrical equipment with the advent of transistorization and microelectronics. Everything became smaller. Japanese companies aimed to produce very pure designs. The company Yamaha turned to the Italian designer Mario Bellini for the *TC-800GL* cassette deck (1974).

The country was experiencing phenomenal growth, as its dazzling exhibits at the 1970 World Fair in Osaka demonstrated. Industrial design was one of the strengths of Japanese industry, as seen in the rise of companies such as Sony, Matsushita, Toshiba and Canon.

Italy continued to be innovative, with Olivetti developing the office sector in the 1960s. Typewriters and computers took on the forms and colours of biomorphic sculptures. Ettore Sottsass produced the *Elea 9003/1* computer (1962) and the Valentine, a red plastic portable typewriter (1969). He designed the *Tekne 3* electric typewriter (1963) with Hans von Klier. Mario Bellini joined the Olivetti team and introduced softer, more ergonomic forms with the *TCV 250* video console (1965), the Logos series of pocket calculators (1970) and the *Divisumma 28* desk calculator (1973), covering the keyboard with a thin, soft rubber film which gave the user a certain tactile pleasure. Rodolfo Bonetto and Naoki Matsunaga produced the *HO1* (1969–70), a digitally controlled machine tool. The Italian company Brionvega, an industrial manufacturer of electrical goods, approached Richard Sapper and Marco Zanuso, who excelled themselves with the very minimal design of their *Black 12 ST 201* television set (1969). Designers were also extraordinarily inventive in the field of lighting. Gino Sarfatti, Mario Bellini and Achille and Pier Giacomo Castiglioni produced designs for the company Flos. Vico Magistretti created the *Éclisse* table lamp (1967), the *Triteti* ceiling light and the *Telegono* lamp, made of resin, while Livio Castiglioni and Gianfranco Frattini designed the *Boalum* light for Artemide. The company Danese manufactured high-quality small objects designed by Bruno Munari and Enzo Mari. They designed many items together, including games and toys; then, on his own, Enzo Mari designed ashtrays and items for the office which were generally simple in shape and ideally suited to mass production. The quality and consistency of these accounted for a great part of Danese's success. The company's aim was to produce objects that were immediately identifiable.

The year 1973 marked the end of this period of euphoria. The oil crisis and the ensuing economic crisis hit companies producing plastic furniture and objects particularly hard. The limits of the consumer society became apparent, and there was no longer any place for the ephemeral and the merely fashionable. Values would have to be reassessed.

Alternative design (1973-81)

- 1973: the energy crisis
- High-tech style
- Anti-design
- Postmodernism and historicism

The happy-go-lucky attitude and freedom of the 1960s were followed by a period of anxiety and instability. The oil shock of 1973 was followed by a feeling of crisis. It was an opportunity for a generation that had now reached adulthood to reassess their values and to start thinking critically about politics, economics and ecology. Academics and sociologists asked questions about the nature and purpose of design. This questioning of design implied a criticism of the consumer society of the 1960s.

1973: the energy crisis

The last embers of the carefree attitude and freedom remaining from the 1960s were extinguished with the oil crisis of 1973. All the same, the Pop culture spirit was still alive and kicking in London on 10 September 1973, when two major shops with totally opposite visions of society opened simultaneously: Biba and Habitat. Biba provided customers with a sophisticated environment which encouraged them to dream – blacks, golds, pinks, mauves and purples, an ambiance of lushness and decadence – designed by Tim Whitmore and Steve Thomas for Barbara Hulanicki. Habitat opened a new shop on the King's Road in London (now called the Conran Shop), more elegant and more expensive than its other branches. It operated on a self-service basis; the goods were straightforward, easily understood and clearly displayed.

The energy crisis in 1973 marked the end of an era. It led to a dramatic rise in the price of oil-based products such as plastics. The whole plastic furniture industry suffered as a consequence, and it became impossible to produce inexpensive furniture. There was a quick reaction from those advocating anti-design and the rationalists, who immediately got a new strategy under way. The international crisis forced everyone with any sort of economic stake to think long and hard about how they could rationalize production methods in order to reduce costs and stay competitive. Designers in the furnishing industry abandoned their creative whims and flights of fancy and directed their efforts towards a more anonymous style of furniture.

Ecological design and responsible design

The crisis led designers to examine their consciences and this was reflected in a return to first principles. The rise in the price of oil hit consumers hard and triggered widespread ecological and humanitarian concerns: there was a rejection of technological constraints, a desire to find solutions to poverty and to help the disabled and the Third World, and a determination to fight against pollution, scarcity of resources and waste. A return to wood and natural materials in general was one of the solutions that was put into effect, after so many years of experimenting with plastics.

Previous page: ***Mezzadro*** **stool, Achille and Pier Giacomo Castiglioni, 1957. This stool is a forerunner of the high-tech approach, and it includes a recycled tractor seat in the spirit of Marcel Duchamp's ready-mades. Manufactured by Zanotta (1983).**

In his book *Design for the Real World* (1971), Victor Papanek (1927–99), a creative designer of Austrian origin, envisaged a new world that would assimilate the idea of revolutionary design – as opposed to the assumptions behind traditional design, based on the economics of buying, consuming and throwing away. Designers would have to recognize the need to reshape tools, the environment and attitudes to the environment, and design would have to remain rooted in practice.

The population increase of the 1970s was nothing compared to the increase in refuse of all kinds. The car was responsible for 60% of atmospheric pollution in the USA. Increased air travel was threatening the planet by contributing to the greenhouse effect.

One of the suggestions for limiting amounts of waste was communal living, whereby people consumed more but owned less. That was the thinking behind the construction of temporary domes – 'cubes' in the manner of Ken Isaacs. At the Drop City commune in Colorado, the dome-shaped houses were built from salvaged materials and all needs were supplied by solar energy. Self-build and standardized elements were developed by the Belgian architect, Lucien Kroll. Given its important ecological and social role, design had to be truly revolutionary and radical: making more with less, manufacturing more durable objects, recycling materials and reducing waste.

In 1976, the ICSID (International Council of Societies of Industrial Design) organized a conference in London on the theme 'Design for Need'. It was the time when the idea of using solar energy for heating was beginning to gain ground. The 'do-it-yourself' spirit was promoted in the book *Nomadic Furniture* and demonstrated in the *Nomadic* range (1973–4) by Victor Papanek and Hennessey. Also in 1976, the first Body Shop opened in London. The products it sold were based on 'green design', which stressed the importance of ecological concerns through the sale of recyclable goods. Designers pledged their commitment to this approach, making choices which affected the very survival of the planet. The architect Richard Buckminster Fuller produced a design for inhabitable structures based on synergetic and energetic geometry. His geodesic structures had been in existence since 1952. In 1963, he launched an inventory of global resources at Southern Illinois University called 'The Era of Survival', but it was not followed up.

In spring 1969, six leading American schools undertook a study into the possibility of creating a living environment in the depths of the ocean. However, it was overshadowed by the publicity given to another project: research into the development of dome-covered colonies on the moon. The mathematician and designer Steve Baer invented solar houses, which he called 'zome works'.

A return to wood

The furniture industry was hit hard by the energy crisis. People started to think seriously about going back to wood and natural materials. Even if less morally committed than the ecologists, industrialists went into reverse without too much grinding of gears. In Italy, Zanotta re-issued Giuseppe Terragni's wooden chair *Follia* (1934), emblematic of Italian rationalist design.

The softer feel of wood also reappeared in the work of Tito Agnoli, who designed the *841* stacking chair, manufactured by Montina. With the *L12* range (1972) for Lema, Angelo Mangiarotti came up with modifiable structures made from natural and painted wood. Furniture in natural colours – black, brown or neutral – was the fashion: the time for bright colours was past. Afra and Tobia Scarpa designed wooden seating for B&B and Molteni. The 14th Milan Furniture Exhibition (1974) hinted at the trend for a return to wood, showing the commitment in that direction of manufacturers such as Boffi, Driade and Bernini. The 15th Furniture Exhibition confirmed the pre-eminence of wood, with a return to the Scandinavian style, the 'teak' look of the 1950s, and Sergio Asti's reinterpretation in wood of a 1920s bed, *Il Lettone*. The *Tomosama* range (1975) in natural pine by Burghard Vogtherr, manufactured in Germany by Rosenthal, also attracted a lot of attention.

Reluctant to get involved in social issues, worried by the oil crisis and frightened by terrorism, the new middle classes sought refuge in a warm, cosy home. In their withdrawal – justified by a 'well-bred' attitude and the enjoyment of nature and culture – they focused on wood as something familiar and secure, turning again to 'luxury' furniture and rejecting 'adaptable' designs. People favoured traditional furniture: armchairs, chairs, tables, chests of drawers, and so on. At the Milan Furniture Exhibition of 1978 there was hardly anything but wood on display. The table became an object for experimentation, as demonstrated by the works of Carlo Scarpa for Bernini, Superstudio for Poltronova, and Mario Bellini for Cassina.

Ergonomics and disability

Designers from the Scandinavian countries were experienced in the use of wood, and they engaged in innovative thinking about everything relating to the world of children, the disabled and working tools. They came up with ecological furniture requiring very few materials, such as tubular metal and fabric. They were inspired by the ergonomic and humanist arguments of Victor Papanek, as set out in his book *Nomadic Furniture* (1973). In Sweden, the designers Johan Huldt and Jan Dranger supported a non-consumerist approach, designing their first items of furniture in cardboard, before founding the company Innovator Design. This produced anonymous designs, and was their way of registering their opposition to 'cult' products. The *Stuns* chair – functional, simple, made from tubular metal and fabric and first shown at the Cologne Furniture Exhibition in 1973 – was hugely successful. The

Table and chair from the *Easy Edges* collection, corrugated cardboard, Frank O Gehry, 1972.

Innovator range was sold through the Ikea chain of stores. The company A&E Design was founded in 1968 by Tom Ahlström and Hans Ehrich, who specialised in living aids (equipment for the disabled and the elderly). Maria Benktzon and Sven Eric Juhlin set up Ergonomi Design Gruppen in Stockholm, studying the muscular mechanisms involved in picking up objects and specializing in design for the disabled.

Finland suffered greatly as a result of the oil crisis. The glass industry, with its high energy requirement, was particularly badly affected. As in neighbouring countries, Finnish designers concentrated on the world of work, on the ergonomics of tools, on safety and on social cohesion. Yrjo Kukkapuro's office chair *Fysio* (1978), manufactured by Avarte, was one of the first office chairs to have a shape dictated by an analysis of the human body. The chair was made of preformed plywood.

The global recession had a deadening effect on Danish design. Only a few leading-edge companies such as Bang & Olufsen and Georg Jensen were able to hold their own at international level.

Recyclables

Faced with the dual predominance of the rational and the superfluous, some designers turned to anti-design. The preoccupations of the anti-design movement in the 1970s were political and social. One of the last designers to be inspired by the Pop movement was the Canadian Frank Gehry (b.1929). He was constantly searching for new uses for those materials, used hitherto in only one fixed way, which had both structural qualities and a good finish. His research into cardboard resulted in the

Easy Edges range of 17 items of furniture made from compressed cardboard (1972); they made use of simple technology, and were recyclable and cheap. The company Jack Brogan made the first models, and in 1982 they were re-issued by Chiru. From his father, who owned a furniture factory, Gehry had learnt to appreciate furniture-making, craftsmanship and a job well done.

But, being first and foremost an architect, he turned to architecture, aware of its social role. He returned to furniture design in the early 1980s with a new collection, *Experimental Edges*, again made from cardboard. This had something of a work of art about it, being produced in a limited edition.

High-tech style

The promoters of high-tech design systematically reassigned industrial and professional objects to domestic use. Deliberately or otherwise, appropriation and collage created a mismatch with reality. The high-tech style was based on the interrelationship of objects taken out of context, one of its principles being to explore the range of objects available and to work out for oneself in what context they might be used. It was a matter of appropriating them and using them in a different context, while preserving their original function (a factory sink, a workshop lamp, a fence, and so on). A great many industrial objects, designed anonymously with no

Omstack chair, Rodney Kinsman, 1972, stacking chair with steel frame and perforated seat and back, manufactured by Bieffeplast. London, Victoria and Albert Museum.

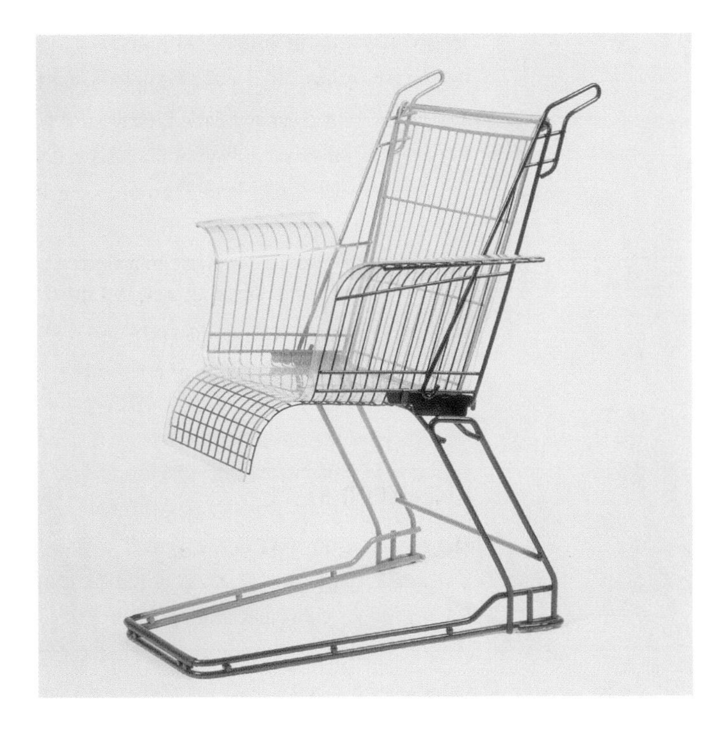

Consumer's Rest armchair, Stiletto, 1983, prototype made in Berlin. Salvaged supermarket trolley, manufactured by Brüder Siegel (1990). Vitra Design Museum.

thought of satisfying the needs of the consumer society, nonetheless turned out to be useful to the general public. They met domestic needs, fitted into the domestic interior, and could be adapted at will. Industrially produced high-tech furniture often had to be assembled by the buyer, but it was cheaper. The high-tech style borrowed its language, materials – glass, iron, steel – and sales techniques from industry.

Appropriation and collage

The Modern Movement, which had always been driven by an endeavour to create a dialogue between art and technology, was put at a disadvantage by the fact that technology was the only factor in the high-tech style. The designer's artistic intervention consisted of appropriating the object and calling it 'art' – a similar process to Marcel Duchamp's use of urinals, bicycle wheels and bottle-racks to produce his ready-mades. Charles Eames was one of the precursors of high tech. He used industrial building elements for his house at Santa Monica in 1949. The Castiglioni brothers used a tractor seat for their *Mezzadro* stool (1957), and a car headlight for their *Toio* lamp (1962). The *Omstack* chair designed by Rodney Kinsman in 1971 is representative of high-tech design. The brief was to produce a low-cost, multi-use chair. Made from perforated sheet steel and available in a wide range of colours, it was intended for both indoor and outdoor use. It was very functional, was stackable and could be clipped together to create rows of seats. The high-tech style appealed to those who had a simple-minded fascination with tools, cold materials and metallic environments as well as a liking for the atmospheres of hospitals, prisons or factories. The punk movement emerged in 1977. The high-tech style fitted in well with the wrecked, mannerless punk world and its contempt for appearances and protocol. It recalled the 'no future' slogan of punk, being an ideology that has no plans – and no roots.

The loft spirit

The American Ward Bennett was the great inspiration for interior designers where the industrial style was concerned. At odds with the Pop style of the 1960s, he turned to industry for inspiration, rediscovering the lamps used in factories, factory fittings, factory shelving and hospital taps. Another designer to promote the industrial style was Joseph Paul d'Urso. For his interior schemes, d'Urso chose cyclone barriers, a rotary dry-cleaning hanger, a surgeon's washbasin, and so on – all articles borrowed from the field of industry. He made use of the full range of features typical of the high-tech style, including mezzanines, scaffolding, angle-irons, and containers on casters. The architect Norman Foster, one of the representatives of the high-tech style in Britain, designed office and factory interiors.

The Pompidou Centre (1977) in Paris gave great prominence to the aesthetics of its structural elements. The opening of this cultural complex – devoted to all fields of artistic activity, but especially the visual arts, musical research, industrial aesthetics, film and the promotion of reading – made a great impression on French design. Occupying 4000 sq m of the Centre Pompidou, the CCI (Centre de Création Industrielle) included a research centre and a publishing house. Documentation classified by designer, manufacturer, distributor, location and price could be consulted there; the information would also find its way into catalogues. In addition, the CCI held exhibitions. Although the state was at last investing in the field of design, this was a far from euphoric period. Many design agencies were hit hard by the repercussions of the energy crisis and had to close. Marketing reasserted its dominance. At the end of the decade there were some private initiatives such as the creation of the VIA (Valorisation de l'Innovation dans l'Ameublement) association in 1979, financed by the furniture industry.

Anti-design

Italy tried to keep up its exports by pursuing its policy of pure design. But during this period of economic instability, a debate began between the adherents of pure design and a minority of avant-garde designers who delivered a sustained criticism of a consumer society which no longer obeyed the laws of supply and demand. They used derision as both a destructive and a creative weapon.

Radical design

The first architectural avant-garde groups, Archizoom and Superstudio, emerged in 1966 from the Faculty of Architecture in Florence. In the unusually intense political debate that was taking place there, architecture was regarded as a political instrument.

Archizoom and Superstudio organized two joint exhibitions, one at the Galleria Jolly 2 in Pistoia in 1966, the other at the municipal art gallery in Modena in 1976. The exhibition 'Superarchitecture' was a theoretical manifesto for Pop architecture. It displayed schemes and prototypes, some of which were put into production by the furniture manufacturer Poltronova. They used absurdity to denounce the way the consumer society had lost its way.

Archizoom was founded in 1966 in Florence by Andrea Branzi, Gilberto Corretti, Paolo Deganello and Massimo Morozzi. The group produced a utopian, derisive concept for a city in the shape of *No-Stop-City* (1970), which used the criteria of spatial organization found in factories and supermarkets, with live-in cupboards. The group also turned its attention to design, producing kitsch furniture: the *Dream Beds*; the *Safari* sofa (1967), made up of elements and manufactured by Poltronova; the *Nepp* chair (1968); and the revolutionary *AEO* seating system (1975), manufactured by Cassina. Another example of this 'design as protest' was the radical interior by Archizoom Associati called *Empty Room* – in which the voice of a little girl could be heard describing the room as a brightly lit, colourful interior. This was presented in 1972 at the exhibi-

The *Imperial Rose Dream Bed*, Archizoom Associati, 1967. The introduction of neo-kitsch into the traditional area of the bedroom. FRAC (Fonds Régional d'Art Contemporain) of the Centre region, France.

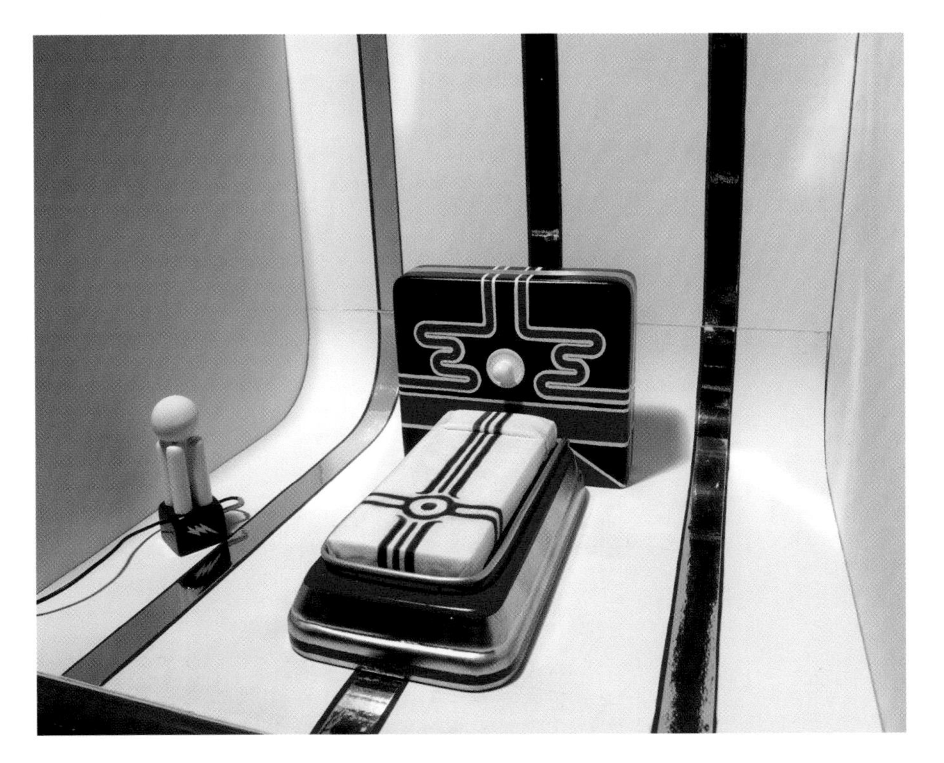

tion 'Italy: the New Domestic Landscape' staged by the Museum of Modern Art in New York. In 1975, the partners in Archizoom Associati went their separate ways, most of them going on to work in Milan. Superstudio, which had already produced its 'histograms of architecture' in 1969, devised a kind of grid made up of simple volumes; it was supposed to be used as a creative aid – by the town planner for towns, by the architect for buildings, and by the designer for furniture.

The group played an important role in the development of the confrontational radical movement in Italy. The members of Superstudio challenged functionalism, accusing it among other things of providing capitalist businesses with publicity arguments under the cover of ergonomic research. Most members of Superstudio became academics, teaching at universities and setting out on the long, silent path of didactic research. They showed their work in the exhibition 'Italy: the New Domestic Landscape' at the Museum of Modern Art in New York in 1972, and in exhibitions in Europe in 1973 and 1974.

A critical attitude

It was in this climate of rebellion that Global Tools was created in Florence in 1973 as a counter-school of architecture and design. This set out to bring together all the groups and individuals of the Italian radical architectural avant-garde and provide them with the same focus. The counter-school was officially founded in November 1974, and an article

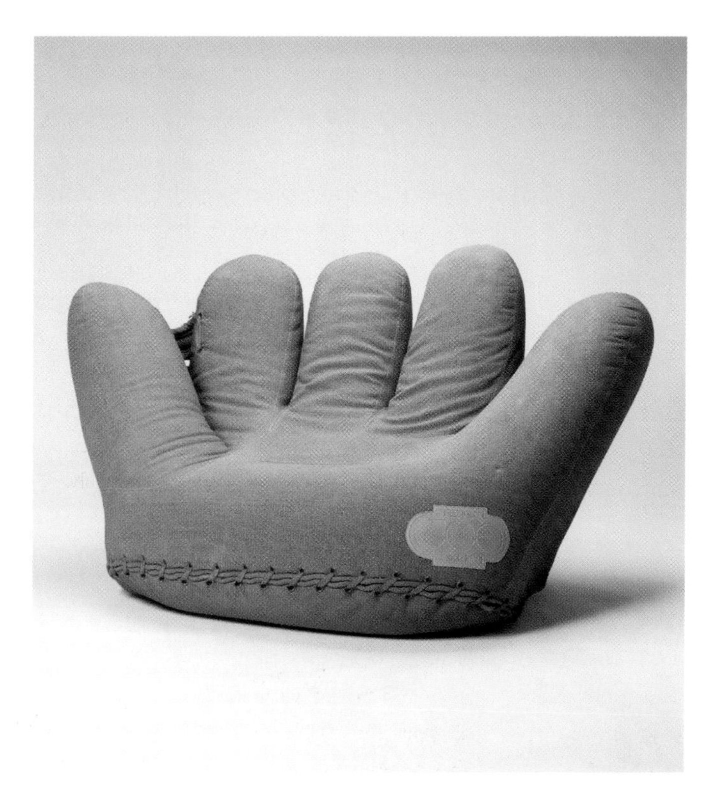

Joe armchair, Jonathan De Pas, Donato D'Urbino, Paolo Lomazzi, 1971. Created in homage to the baseball player Joe DiMaggio. The seat of the leather chair is shaped like an oversized glove. Vitra Design Museum.

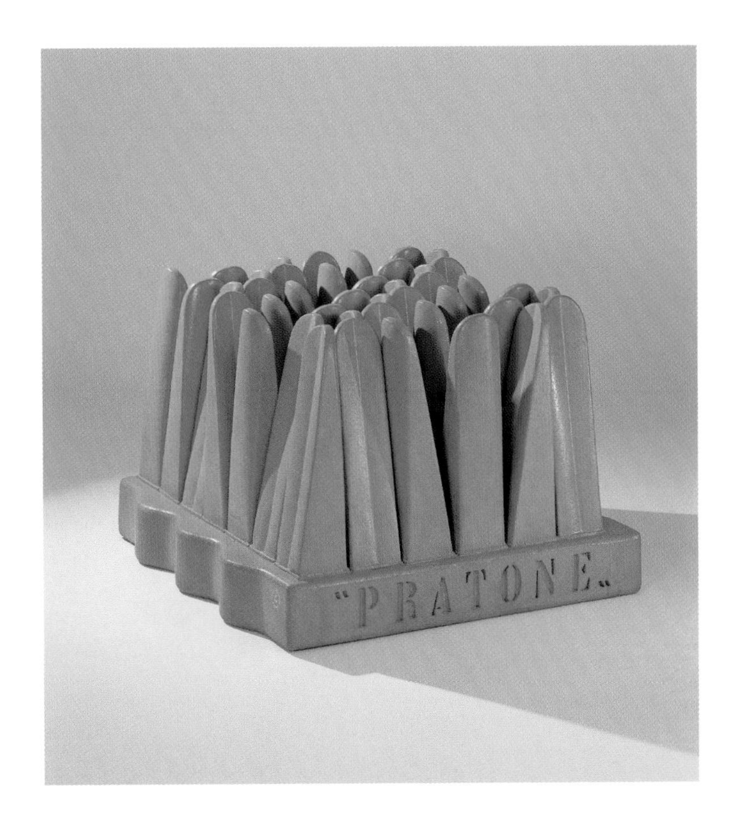

Pratone ('Big Meadow'), Gruppo Strum (Giorgio Ceretti, Piero Derossi, Ricardo Rossi), late 1960s, seat-cum-sculpture manufactured by the company Gufram, in their Multipli collection. Vitra Design Museum.
Vitra Design Museum.

in the magazine *Casabella* in January 1975 set the seal on its recognition. Teachers and pupils joined together to form a single fighting front with a strong sense of solidarity. However, the institution did not last beyond 1975, marking the end of the first phase of anti-design. Political and social problems resulting from the energy crisis coupled with the threat of terrorism no longer left any room for critical discourse. All the same, some groups – such as UFO, 9999, Studio 65 and Gruppo Strum – adopted a radical stance.

The work they produced was not made available to the general public, but it was well received by the press. Italian design was influenced by art, and by Pop Art in particular, with references to the work of Claes Oldenburg. Jonathan De Pas, Donato d'Urbino and Paolo Lomazzi actually designed a chair in the shape of an oversized baseball glove and named it *Joe* after Joe DiMaggio. The *Bocca* sofa by Studio 65, manufactured by Gufram, was a reference to Salvador Dalí and his *Mae West Lips* sofa.

Land Art was another source of inspiration, seen in the output of the Italian manufacturer Gufram, the foam-rubber multiples created by artists, the giant stones *Sassi* and *Sedilsasso* (1970) and the *Pavepiuma* floor covering in the form of pebbles (1970) by Piero Gilardi. The giant-sized imitation of grass, also in foam rubber, in *Pratone* (1966–70) by the Gruppo Strum is reminiscent of Giuseppe Penone's use of organic materials. In its shape and in the material it conjures up, the *La Cova* sofa

Capitello ('Capital'), 1971, armchair in polyurethane foam, covered with Guflac elastic paint, Studio 65. Vitra Design Museum.

(1973, 'The Nest') by Paolo Ruffi brings to mind some of Mario Merz's *Igloos*. Even the reference to antiquity in *Capitello* (1973) by Studio 65 is reminiscent of the Graeco-Roman allusions in some works by Giulio Paolini. Manufactured in limited runs, these products would prove to be as important in the history of design as the most successful mass-produced objects.

Alchimia, 'a milestone in the history of design'

Alchimia was a group of designers who had a major influence on all sectors of design: industrial design, interior design, graphic design, fashions and product design.

Their work sits somewhere between modernity and postmodernity. Formed in Milan in 1976 by Alessandro Guerriero (b.1943) and his sister Adriana, the Alchimia group was well known for its experiments with radical design. Initially it was a graphic design studio. Alessandro Guerriero met Alessandro Mendini (b.1931), the chief editor of the journals *Casabella*, *Modo* and *Domus*, at an exhibition on 'radical suitcases' at the Alchimia studio (the theme of the exhibition related to the interpretation of the suitcase). Guerriero helped Mendini to create items of furniture for an exhibition in June 1978 at the Palazzo dei Diamanti in Ferrara, where Mendini exhibited both the famous *Proust* armchair and the *Kandinsky* sofa; Ettore Sottsass and Andrea Branzi were also among the exhibitors. In 1978, the group became the experimental centre of the

Milanese radical tendency, combining crafts and industrial processes. In order to preserve its autonomy, it opened a gallery to show its own work, including its *Bauhaus* range. It was at the 1979 Furniture Exhibition in Milan that Alchimia made its first big impact. An article in the magazine *Casa Vogue* called their contribution 'a milestone in the history of design', a real break with standard design. The group came up with anti-rational design products, using historical references and decorative motifs from the 1950s. They revisited design classics in a humorous way. They employed references to art while at the same time trivializing it. The pointillism of the painter Paul Signac was applied to a copy of a Louis XV armchair, resulting in Alessandro Mendini's *Proust* armchair.

The company Abet Laminati manufactured Paola Navone's *Gadames* dressing-table and the *Le Strutture Tremano* table by Ettore Sottsass Jr, with its slender, unsteady legs, laminated wooden base and glass top.

The design language used by the Alchimia studio was ironic in tone, rejecting rationalist discourse. It can be seen as a continuation of the radical movement of which Germano Celant was the theorist. Alchimia did not always look to produce novel designs: they also recycled and decorated. Their work was based on form or pictorial design. 'Re-design' became a postmodern concept. The members of the group sketched out

Proust armchair, Alessandro Mendini, 1978, for the Alchimia studio. The armchair, inspired by the Louis XV style, is hand-painted in the manner of the pointillist painter Signac.

projects which were then produced in small numbers, without any concern for productivity. Alchimia began to be affected by changing moods and sentimental design. The group ran into serious difficulties when the designers wanted their designs to go beyond the experimental prototype stage and into production. Guerriero was more interested in exhibitions and cultural activities than in marketing. In the end the nihilistic, critical attitude of Mendini prevailed, and the Alchimia studio gradually became identified with his ideas. The break-up was gradual. Ettore Sottsass left the group in autumn 1980, followed some time later by Michele de Lucchi.

Memphis

This group was founded in 1981 at the instigation of Ettore Sottsass, working with Michele De Lucchi. The young Milanese architects and designers Fausto Celati, Ernesto Gimondi, Renzo Brugola, and Mario and Brunella Godani later joined the group. Memphis produced its first designs in February 1981. Lights were made (by Artemide), furniture was manufactured, ceramic objects were produced, and on 18 September 1981 their first exhibition – 31 pieces of furniture, 3 clocks, 10 lights and 11 ceramic objects – opened before an audience of 2,500 people. George Sowden, Marco Zanini, Michele De Lucchi, Matheo Thun and Nathalie du

Carlton shelving, Ettore Sottsass, 1981, produced in small runs by the Memphis studio. A new demonstration of craftsmanship.

Ettore Sottsass

Dining chair made of tubular metal and wood covered with patterned melamine, Ettore Sottsass in collaboration with the company Abet Laminati, 1980, *Bauhaus* collection, Alchimia.

Sottsass, an Italian architect and designer of Austrian origin, was born in 1917 at Innsbruck. After studying architecture at the Politecnico in Turin until 1939, he had to wait for the end of the war before setting up as an architect. He played a part in the post-war reconstruction of Turin and Milan. He was also interested in design and painting, exhibiting at the Milan Triennale and showing his more personal works at the Galleria del Naviglio in Milan in 1956. He quickly became known as one of the leading Italian representatives of rationalism and functionalism in design and architecture. He took a particular interest in redefining the domestic living space, seeking a subtle balance between structure (with its intellectual rigour) and colour (seen as a source of energy and vitality). In 1959, he became a consultant designer for Olivetti. Trips to the USA and India altered the way he saw the world and he began to appreciate the close relationship between man and nature. The culture of the East served as inspiration for his ceramic works.

Interested in 'pure research', he sought to transform the conventional living space. Furniture was harmonized with walls through the use of coordinated colours. In 1970, the year of the *Grey Room*, he used synthetic materials. In 1972, he took part in the exhibition 'Italy: the New Domestic Landscape' at the Museum of Modern Art in New York, where he showed a 'habitat profile' consisting of movable plastic modules. In 1976, he was invited by Hans Hollein to take part in 'Man Transforms', the opening exhibition of the Cooper-Hewitt Museum in New York. That same year, the International Design Centre in Berlin organized a retrospective exhibition of his work, which was subsequently shown at the Venice Biennale and in several other cities. In 1977, Sottsass was involved in a project for creating the signs at Fumicino Airport in Rome. From 1978 to 1980, he was part of the Alchimia group. In 1980, he founded the company Sottsass Associati with Aldo Cibic, Matteo Thun and Marco Zanini. That same year, he launched the Memphis group, for which he created a large number of objects and items of furniture.

Gaetano Pesce

Green Street chair, Gaetano Pesce, 1984, polyester, steel and foam chair, manufactured by Vitra International. Saint-Étienne, Musée d'Art Moderne

Pesce was born in La Spezia in Italy in 1939, and attended the Faculty of Architecture in Venice. Interested in figurative art, he exhibited at various Italian galleries. He also studied at the Institute of Industrial Design in Venice. In 1959, he was involved in founding the Gruppo N in Padua. As a member of Gruppo N, he collaborated with the Gruppo Zero in Germany, the Groupe de Recherche d'Art Visuel (GRAV) in Paris, and the Gruppo T in Milan. He had many contacts at the Ulm School of Design, where he exhibited. He gave many lectures and presented his *First Manifesto for an Elastic Architecture* (1961). In 1965, he met Cesare Cassina, designing the *Yeti* armchair for the company Cassina in 1968. Pesce founded the company Bracciodiferro with Francesco Binfare in order to manufacture experimental objects. He designed the *Up* series of chairs, made from polyurethane foam and covered with jersey, for the company C&B. In 1971, he put forward a scheme for a living unit for two people that was part of an overall project for an 'underground city at a time of major contamination' at the 'Italy: the New Domestic Landscape' exhibition in New York in 1972. In 1975, the Musée des Arts Décoratifs in Paris devoted an exhibition to his work: 'Le Futur Est Peut-être Passé', organized by the CCI.

The research carried out by Pesce became a benchmark for new Italian design. It suggested that new technologies for the manufacture of objects should be introduced and broke with the repetitiveness of mass production, as is demonstrated by the very different series of chairs he designed: the *Golgotha* chair (1972), the *Sit Down* armchair (1980) and the *Dalila* chair (1980), all manufactured by the furniture company Cassina. Pesce then produced examples of accidental design, in which manufacturing defects were made into features: for example, the *Sansone 1* table (1980) and the *Pratt* chair (1983). In 1986, the *Feltri* armchair involved the appropriation of felt, an industrial material. Pesce has taken part in many schemes and competitions: a 'vertical loft', a skyscraper in Manhattan (1978), the Chicago Tribune Building in New York (1980) and the redevelopment of Les Halles in Paris (1979).

Pasquier produced furniture, lighting, silverware and ceramics. The exhibition enjoyed great success and the project was developed commercially thanks to the support and collaboration of four partners who founded a company to distribute Memphis's designs on the international market. These four partners were the furniture manufacturer Renzo Brugola; the lighting manufacturer Fausto Celati; Ernesto Gismondi, the chairman of Artemide; and Brunella Godani, the owner of a gallery in Milan which became Memphis's exhibition space. Barbara Radice was responsible for the artistic direction and the coordination of the group, while Ettore Sottsass was its central figure. He put across his his ideas as though he were engaged in a large-scale cultural campaign. Memphis immediately opened its doors to international designers, especially those great names of the postmodern movement Michael Graves and Hans Hollein.
Young designers of every nationality – some unknown, and some, like Arata Isozaki or Shiro Kuramata, with a certain reputation – were brought together at this first exhibition. The success of this collection was confirmed by coverage in the international press.

At the Milan Furniture Fair in 1981, the press marvelled at the brightness of the Memphis colours, the basic forms of the furniture and the diversity of the objects on show. One of the group's innovations was the use of laminated plastic. This covering material, generally found in kitchens and bathrooms, was freely used by Memphis for its intrinsic qualities: its coloured, sometimes patterned, surfaces served to decorate furniture in brightly lit sitting-rooms. The company Abet Print used its expertise to help the group realize its decorative fantasies. Memphis designers – especially Sottsass, Sowden and De Lucchi – were wary of the 'design' label and stressed the decorative and ornamental aspect of their work. The decorative qualities of Memphis are to be found in a deliberate avoidance of cultural references and great figurative freedom.

Postmodernism and historicism

The postmodern movement involves rejection and discontinuation, rather than a deliberate choice. It began its reaction against modernism as early as 1968. The roots of the postmodern rebellion lay in a realization of the radical changes taking place in social relations, production methods, and industry. The crisis over energy resources led to the re-emergence of problems that people thought had been solved long ago. Rather than follow an ultra-futuristic direction, the postmodern movement turned back to the origins of design and tradition, and attempted to combine present and past. People's behaviour, their clothes, revival jazz and folk music – all displayed the same trend. The need for contemplation and communion with nature, far away from machines, began to make itself felt.

Tea and coffee service (Tea and Coffee Piazza collection), Aldo Rossi, 1983, made by Officina Alessi.

A Renaissance of sorts

The word 'postmodern' infiltrated, with mixed fortunes, different areas of the humanities: literature, semiology, philosophy and architecture, where it moved on from critical theory to affect architectural practice. The anti-design postmodern movement reached its peak in architecture. To judge from the urban landscape, modern architecture seemed to lack any vital quality, to ignore collective values and local characteristics, and to be dreadfully uniform. The architects of the postmodern movement focused on the city with a view to renewing it at all costs and disrupting the existing balanced state of things; but they would integrate the rules they had been taught and their knowledge of the canons that had governed centuries of history. In this way they drew attention to a place's heritage. It was a Renaissance of sorts, setting out to recover some of the values of the past.

Postmodernists incurred the wrath of the guardians of modernity by arguing the case for the rooting of postmodern theories in history, and by contrasting this with the alliance between modernity and bureaucracy and totalitarianism. They proclaimed their determination to break away from a movement obsessed with technological innovation. The architect Charles Jencks set out to elucidate the specific nature of postmodernism by defining the language of postmodern architects as an element of communication. The language of a postmodern building is the language of archetypes, taking architectural conventions as its source. This may be illustrated, for example, by the reuse of a classical Greek temple: architecture returning to its core history.

Architecture and design had been trying to escape from the Modern Movement since the 1950s. Postmodernism also prompted new thinking about town planning and unbridled development. The energy crisis encouraged a search for alternatives to excessive urban growth and led to the choice of more traditional materials.

The American architect Robert Venturi was one of the advocates of this approach. He rejected the purist, reductionist attitude of modernism and championed the values of 19th-century America with the Wislocki House and the Trubek House in Massachusetts (1970). He liked urban kitsch, neon signs and advertisements. Philip Johnson, another American architect, placed an unusual pediment on his AT&T Building (1978–82) in New York. For the Piazza d'Italia (1974–8), which he built in New Orleans, Charles Moore used columns, capitals and fluting. Aldo Rossi stacked up simple forms in the manner of a child's building blocks. Hans Hollein structured his projects like antique spaces. Michael Graves, Aldo Rossi and the artists of the postmodern movement tried to move design in the direction of fine art rather than in the direction of industry. Focusing on furniture and objects, they created a style full of imagination and colours which fitted in well with postmodernism. It also fitted in well with the new lifestyle based on a revival of consumerism.

A renewed interest in tableware

The Italian company Alessi, founded in 1921, has its headquarters in northern Italy in the small town of Crusinallo. It makes high-quality metal tableware. In 1982, it created a new brand, Officina Alessi, which produced not only re-issues of historic objects designed by Josef Hoffmann, Christopher Dresser and Marianne Brandt, but postmodern designs as well. It approached internationally renowned architects – Aldo Rossi, Robert Venturi, Michael Graves, Arata Isozaki, Richard Meier, Paolo Portoghesi, Stanley Tigerman, Hans Hollein and Charles Jencks – and asked them to design a tea and coffee service. Michael Graves designed a six-piece service made of silver, aluminium, imitation ivory and Bakelite. This was micro-architecture for the table, a refined amalgam of Biedermeier, Wiener Werkstätte and Art Deco styles. It was the first time that Michael Graves had turned his attention to design; up to then he had created only showrooms. As a result of its innovative policy, Alessi enjoyed huge commercial success, and there was a craze for tableware – something that had not happened for a long time. Designers focused on tableware and silverware, and the objects they produced were like tiny examples of architecture. Objects produced by Alessi – real icons in the early 1980s – were made using techniques that were both craft-based and highly sophisticated.

Re-issues

In 1977, the company Cassina began re-issuing furniture designed by masters of the Modern Movement: Mackintosh, Rietveld, Le Corbusier and Asplund. In 1972, the company Zanotta re-issued the *Follia* chair (1934), designed by Giuseppe Terragni, and in 1976 the company Bernini decided to manufacture *1934*, dining-room furniture by Carlo Scarpa (1934). A 'retro' trend swept through the 14th Furniture Exhibition in Milan. In France in 1978, Andrée Putman created Écart International, which re-issued historic furniture: Robert Mallet-Stevens's chair, Mariano Fortuny's lamp, Jean-Michel Frank's armchair, and Pierre Chareau's coffee table. This urge to revisit the great classics of design history provided a welcome breathing space after the period of euphoria in the 1960s and the formal digressions of 'sculptural' design.

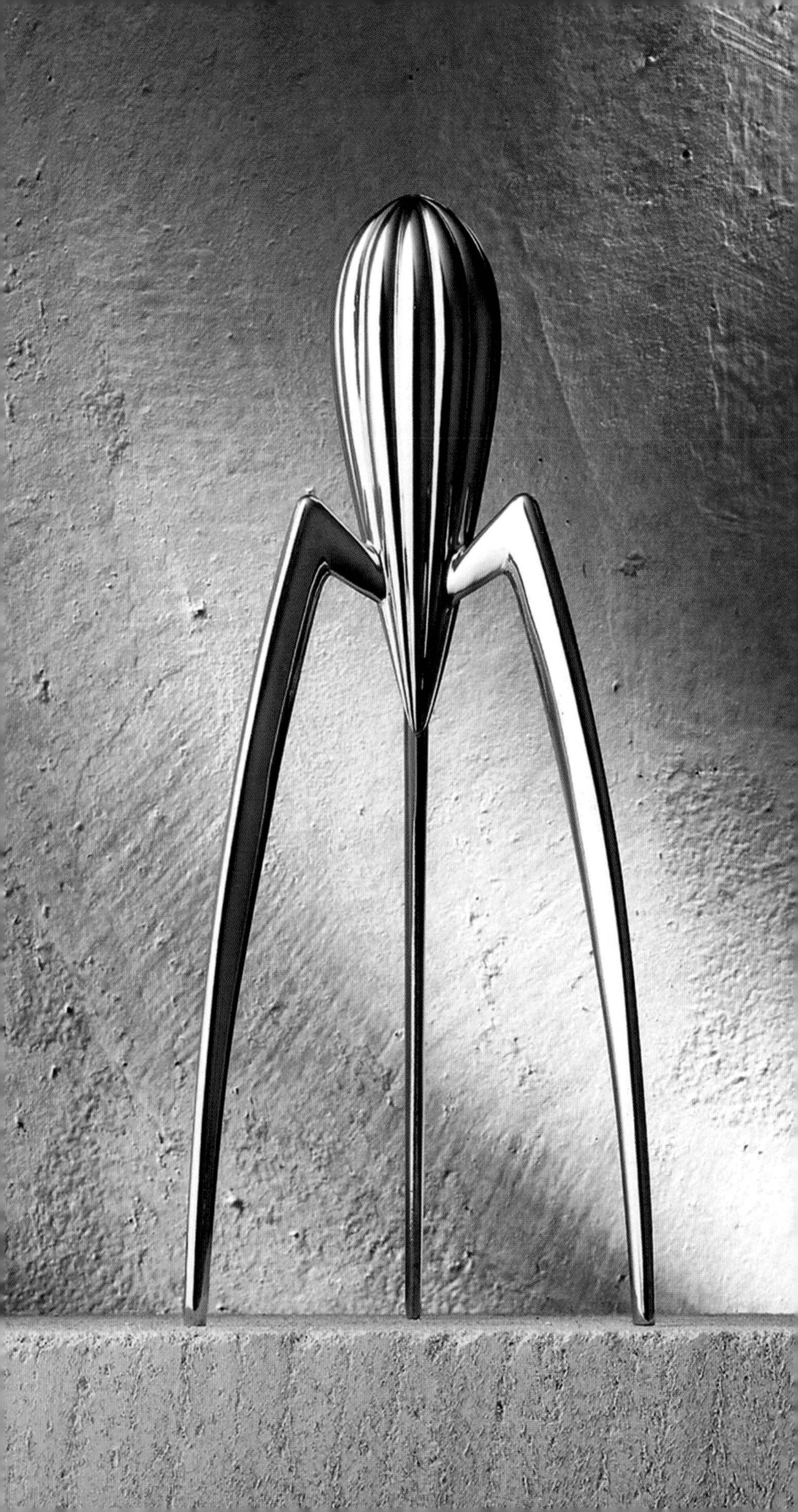

The 1980s: the eclectic years

- French individualism
- Japanese tradition
- Italian contrast
- Britain in the 1980s
- Avant-garde design

After a long period of austerity and recession, the 1980s brought hope, thanks to an economic boom and the end of the Cold War, symbolized by the fall of the Berlin Wall in 1989. But the decade was also associated with the spread of Aids, unemployment and the discovery of a new category of citizen: the homeless. Against a background of seeming disorder, this period was a stimulating one for design, which was boosted by the easy availability of speculative money.

French individualism

Why did French design struggle so much in the 1980s? Was it simply that consumers were too conservative? Young city-dwellers were not interested in accumulating assets. Ever since the 1960s they had opted for expendable furniture that they replaced as soon as it wore out. French industrialists were disinclined to take risks. They refused to accept that design expresses a message, and that furniture, over and above its physical appearance, functions as a sign system indicative of a specific period.

Independent operators

The 1980s saw the emergence of many methods of manufacturing and selling that broke with the industrial norm. The time of 'creative' artists and designers had arrived. Works by sculptors, designers and interior designers were produced in limited editions. Gallery owners invested in this work, trying to anticipate the future development of the market. The gallery which pioneered this approach – the Galerie Neotu on the Rue du Renard in Paris – staged the exhibition 'Onze Lampes' in January 1985. This gallery was founded by Pierre Staudenmeyer and Gérard Dalmon, who both collected objects by the Memphis group. The first people they invited to exhibit included the designers François Bauchet, Dan Friedman, Élisabeth Garouste, Mattia Bonetti, Kristian Gavoille, Éric Jourdan, Pucci de Rossi, Sylvain Dubuisson, Borek Sipek and Martin Szekely. Their choice was eclectic, but governed by the desire to show that design objects could be works of art. Early in 1986, the antiques dealer Yves Gastou became interested in radical Italian design and invited Ettore Sottsass to design the interior of his gallery; later, he also asked Ron Arad, in 1986, and Shiro Kuramata. The shop 'En Attendant les Barbares' was a source of entertainment for journalists, presenting works by Migeon-Migeon, Éric Schmitt, Jarrige and, later, Cheriff; the Barbarian movement eventually emerged from this. Marco de Gueltz, André Dubreuil, Javier Mariscal and Mark Brazier-Jones had their works displayed in the shop 'Avant-Scène' on the Place de l'Odéon.

Previous page: ***Juicy Salif*** **lemon squeezer, Philippe Starck, 1990. Manufactured by Alessi.**

State patronage

The ARC (Atelier de Recherche et de Création) – the research and creative

Suite of office furniture for the French Ministry of Culture, Andrée Putman, 1984.

design workshop of the Mobilier National (the government department in charge of state-owned furniture) – was set up in 1964 with the aim of supporting the transformation of French infrastructure such as local cultural centres and administrative offices in new towns.

Its mission was to encourage experimental research and promote innovative projects: for example, César's research into expanded foam in 1968, or the definition of a four-roomed flat by Olivier Mourgue. After the 1973 oil crisis there was a move back to traditional materials, which reinforced a tendency towards conservatism. The ARC reflected the contradictions of a period caught between 'design' and 'furniture by artists'. With the emergence of the Memphis movement in October 1981 – an Italian group, founded in Milan at the instigation of Ettore Sottsass, which was dedicated to the design of objects and furniture and which aimed to be controversial – design entered 'the world of the image'. The work produced by Memphis was eclectic, and the term 'creator' was used in preference to the term 'designer'.

The 1980s saw a resurgence of public commissions. At his first press conference in September 1981, the French President François Mitterrand announced his policy of 'major public works', mainly overseen by the Minister for Culture, Jack Lang. In 1982, five designers were selected to renovate the President's private apartments at the Élysée Palace: Ronald Cecil Sportes, Annie Tribel, Jean-Michel Wilmotte, Philippe Starck and Marc Held. In his choice of five interior designers, President Mitterrand signalled that he wanted to encourage the creation of original pieces of furniture.

He also wanted pieces that reflected the range of significant trends in French furniture at the time. Pluralism was one of the fundamental ideas of the period.

Jean-Michel Wilmotte was also asked to design the interior of the office of the French Ambassador to Washington (1984), while Isabelle Hebey designed the interior of Mme Mitterrand's office on the ground floor of the Élysée Palace. Work began on the first major projects: the Grand Louvre (enlargement of the Louvre) and the new Ministry of Finance at Bercy. Jack Lang commissioned the furniture for his office from Sylvain Dubuisson (1991), while Henri Nallet, the Minister for Agriculture, chose furniture made to designs by Richard Peduzzi (1989), and Claude Évin, the Minister for Solidarity, turned to Marie-Christine Dorner (1991).

The VIA (Valorisation de l'Innovation en Ameublement) was a French association created on 18 April 1980 under the joint aegis of the Ministry for Industry and Unifa (Union Nationale des Industries Françaises de l'Ameublement). Its director, the designer Jean-Claude Maugirard, was instructed to assemble a committee of experts – industrialists, distributors, creative designers, journalists and institutional investors – who would be responsible for ensuring the successful implementation of any action likely to promote innovation in French furnishing. Several tools were put in place; these included the 'appels permanents' system – whereby a proposal could be put forward at any time and, if approved at one of the two-monthly committee meetings, developed by the designer into a full-size prototype – and the 'Carte Blanche', a kind of bursary granted to designers to use as they saw fit. In order to encourage creative design, the state also organized a competition for office furniture and another for lighting, with government commissions as prizes.

After Memphis

The great merit of the Milanese group Memphis was that it had cast off the shackles of design, disowning any moral ideology it might lay claim to. Memphis made kitsch and a polyglot culture attractive to many people. The movement appealed especially to a certain fringe group in the French population: young, ambitious, upwardly mobile executives. They appreciated whimsical, humorous design rather than design that was functional. The eclectic output of the movement was mainly concentrated in the field of furniture, for it was simpler to give symbolic value to a chair than to a machine-tool or a domestic electrical appliance.

Barbarian chair made from wrought iron and animal skin, Élisabeth Garouste and Mattia Bonetti, 1985. Paris, CCI-Georges Pompidou Centre/Kandinsky Library.

Charismatic creative designers kept well away from big industry, sometimes producing furniture-cum-sculpture in limited editions. Such designers kept the design process close to the level of experiment and prototype, seeing it as their role to create products through formal or stylistic innovation.

François Bauchet (b.1948), a sculptor by training, devoted most of his time to producing furniture that looked impressive but had no obvious functional purpose: the *Epiploon* armchair (1984–7), *ADL* (1986), the *APF* writing desk (1986), and the *Coiffeuse* (1981). In 1989, he created a range of furniture made from resin. Most of these designs were produced by the Galerie Neotu.

The design studio Totem was set up in Lyons in 1980 by Frédéric du Chayla, Jacques Bonnot, Vincent Lemarchands and Claire Olivès. It exhibited its first prototypes in Lyons in 1981, then at the VIA in Paris. Its eclectic, playful furniture is made of wood painted in bright colours: the *Lolypop* chair, the *Chameleon* armchair and the *Zig-Zag* table.

Olivier Gagnère (b.1952) worked with the Memphis group in 1980–1. In 1983, he became interested in small objects, producing them in a style that was magical, even mystical. Impressed by the ceremonial of the Church, he used it as a source of inspiration for his votive lamps, chalices, altars and mirrors: the *Lampe Autel* (1985), *Barbiere* mirror (1987), and *Commode* (1987) were produced by Adrien Maeght. Gagnère was awarded a Carte Blanche bursary by the VIA in 1982. His furniture collection *Banc* (1988), which included a bench and a coffee table, was manufactured by Artelano.

Élisabeth Garouste (b.1949) and Mattia Bonetti (b.1953) studied the ancient Roman style with its drapes and stucco work, which they used for the decoration of the Privilège Club, at the Palace in Paris.

They then decided to start again by going back to the dawn of civilization and prehistory: to blocks of rock and pieces of wood. Their work, influenced by primitive civilizations and at odds with industrial design, led to a trend known as 'Barbarian'. Pucci de Rossi juxtaposed materials which evoked contrary feelings: for example, wood and steel. Many young designers adopted this style, which was marked by humour and irony. With *Lampe de Bureau* (1986), *Stèle Tournant* (1984) and *Armoire Menhir* (1987), the duo known as Bécheau-Bourgeois (both born in 1955) engaged in research that focused on lightness, transparency and mobility, using foams and plastic-covered metal sheets. Pierre Charpin (b.1962) also explored the constituent elements of design. The two architects in the Nemo group – Alain Domingo (b.1952) and François Scali (b.1951) – questioned function. They displayed a great sense of irony and typical malice in their design for the *Moreno and Marini* armchairs (1986), based on the idea of the couple. Coll-Part, a designer of extremes, produced objects aged or spoilt by the passage of time: *Un Jour la Terre Trembla en Mal* (1987, VIA).

More austere design

Towards the mid-1980s, economic realism and austerity produced a particularly trendy style, derived from the Japanese Yohji Yamamoto, who adopted black as his trademark colour. Black epoxy tubing was the component of choice in this new rigorous approach. Sylvain Dubuisson (b.1946) became interested in rare materials. In 1987, he was the first designer to exploit the properties of composite materials, producing the *Composite* table, made from carbon-fibre taffeta. In 1990, thanks to a Carte Blanche bursary from the VIA, he designed the *Aéro* armchair, the *Table Portefeuille* and the *Composite* chairs, as well as many interiors. Martin Szekely (b.1956) was also interested in materials such as Corian (stainless steel). In 1981, after being awarded one of the first Carte Blanche bursaries by the VIA, he came up with the *Soft Foot* range (1981) and the *Pi* chaise longue (1982), complete with table and pedestal table. This marked the beginning of a collaboration between

Philippe Starck

Louis XX armchair, Philippe Starck, 1994: humour and irony. Manufactured by Vitra. Vitra Design Museum.

Philippe Starck was born in Paris in 1949, and trained at the Camondo School. He was initially interested in seating, designing the *Bloodmoney* armchair (1977), the *Francesca Spanish* folding chair (1979–80) and the *Mr Bliss* seat (1982). These marked the start of his commercial success. He carried out a few interior design projects, and created a number of lights – *Easylight* (1980) and *Stanton Mick* (1979) – and the *Tippy Jackson* table, manufactured by Driade (1981). As the result of a Carte Blanche bursary by the VIA, he designed *Miss Dorn*, a chair made of epoxy resin, tubular steel, leather and fabric (1983). The year 1984 was important for Philippe Starck: he designed the interior of the Café Costes in Paris, and gained recognition from the media and the public.

As artistic director of the manufacturer XO, he reclaimed the idea of service to the user and focused on functionality at every stage of the process: production, distribution and use. The *Richard III* club armchair (1984) and the *Docteur Sonderbar* armchair (1985) were manufactured by XO. In 1985, Starck adapted to the demands of mass distribution with articles in folding kit form. He worked with the Les 3 Suisses catalogue company and, unlike other designers, showed an interest in mail-order. For Driade he produced two ranges of wooden chairs: *Anna Rustica* (1986), and *Bob Dubois* (1987). The *Dr Glob* chair (1988) is a modern tubular-steel stacking chair, manufactured by Kartell. Then his forms became more fluid; he moulded aluminium before exploring polyurethane with the *Louis XX* chair (1992), manufactured by Vitra. Among the many objects he has designed are *Walter Wayles II* clock (1987), the *Juicy Salif* lemon squeezer (1987) and the *Max le Chinois* sieve (1990), all manufactured by Officina Alessi. He has continued his involvement in industrial production with mass-produced objects – in particular a toothbrush for Fluocaril in 1992 – while at the same time starting to think about nature and how best to conserve it.

Atlantique high-speed train, Roger Tallon, 1999. The *Duplex 231* train is the first in the 34 series ordered by the SNCF (French Railways). Roger Tallon has collaborated with the SNCF ever since he designed the first high-speed train model, the *001*.

Szekely and the Galerie Neotu, which showed his work and produced the *Containers* series (1987). He made disproportionately tall furniture – *Haut à Rideau* (1987) – and went on to collaborate with many other companies, for example with the Gien, Swaroski and Delvaux potteries. In 1996, he designed the Perrier bottle.

The architect Jean Nouvel (b.1945) was also interested in furniture and furniture-making techniques. In 1987, in receipt of a VIA Carte Blanche award, he designed a collection of metal furniture: the *PTL* positioning table, the *AAV* extendible bookcase, and the *BAO* chest (functioning on the same principle as a tool box). In 1992, he produced an 'office equipment kit' comprising tables and cable trunking units for the CLM/BBDO advertising agency, then the *Less* series for the Cartier Foundation, both manufactured by Unifor.

Industry

Ever since the Industrial Revolution, art, architecture and design have generally been attuned to developments in science and technology. Every new advance provokes a reaction in the creative world as the new material is adopted or appropriated.

The development of carbon fibre and plastics marked a new beginning. For the first time human beings were inventing materials. Faced with new technologies – the microchip, fibre optics, computer networks – and the blurring of the distinction between space and time, designers and creative artists were at a loss. Some large companies had design offices which came up with ways of adapting to these developments. Tim Thom, a team of designers led by Philippe Starck, worked for the multimedia company Thomson. The car company Renault was given a considerable boost by the industrial design unit founded by Patrick Le Quément, who was awarded the National Grand Prix for industrial design in 1992.

On 27 September 1981, the first high-speed train route linking Paris to Lyons was officially opened, with trains travelling at over 300 kph. The trains were designed for the SNCF (French Railways) by Roger Tallon, who carried out research into basic forms and aerodynamic aesthetics (1972–90).

Japanese tradition

Japan emerged onto the international design scene in 1970, when it hosted a World Fair for the first time, in Osaka. The fair was firmly anchored in Japanese tradition.

The Festival Plaza was designed in three dimensions by the architect Kenzo Tange, with a translucent tent roof 30 metres above the ground. It was overlooked by Taro Okamoto's Tower of the Sun, where there were two robots moving around, designed by Arata Isozaki.

Design objects versus new design

An outstanding representative of new Japanese design was Shiro Kuramata (1934–91); his curvilinear storage units, *Side 1* and *Side 2*, manufactured by Cappellini, were shown at the Milan Furniture Exhibition in 1985. He brought out the structural complexity of objects, as in the *How High the Moon* armchair, made from a lattice of nickel-plated steel (1986), as well as the poetry of objects, as in the *Miss Blanche* acrylic chair (1988). The approach of Toshiyuki Kita (b.1942) is full of humour and originality, as demonstrated in the *Wink Chair* chaise longue (1980) with its movable, coloured headrests and the *Kick* table (1983). A year after they had come out, these models were acquired by the Museum of Modern Art in New York. Two other Japanese designers, Masaki Morita and Takamachi Ito, referred back to the Japanese graphic tradition in their work.

Unlike the craft-based production of expensive 'design objects' (often elevated to the rank of works of art and sold in galleries), 'new design' worked with industry and enjoyed great commercial success. The company Fuji overturned established habits by launching the 24-exposure disposable camera in 1986. Canon made photography still more accessible by bringing in automatic computerized focusing on 24x36 cameras, with the *T90* model (1986). Amateur photographers could now shoot their own films with 8mm video cameras. In the field of two-wheeled vehicles, Japan was the global leader with Honda (the world's number-one motorcycle maker), Yamaha, Suzuki and Kawasaki. The fashion designer Kenzo caused surprise by moving to Paris in the late

Tizio lamp made of metal and synthetic resin, Richard Sapper, 1972. Manufactured by Artemide.

1960s. In Japan, an aesthetic approach close to the kabuki tradition emerged, represented by the fashion designer Yohji Yamamoto. Issey Miyake also used the tradition of Japanese dress in his designs. Working as a design stylist for the brand Comme des Garçons, Rei Kawakubo exhibited all-metal chairs – *Comme des Garçons No.1* – at the Milan Furniture Exhibition in 1986. His design ideas kept faith with his overall concept of a universe characterized by the colour black and cold materials: an empty, sanitized universe.

Italian contrast

In 1980, Italy was the world's number-one furniture exporter, ahead of Germany, in spite of the recession. The 31st International Design Conference, held at Aspen in Colorado, had 'The Italian Idea' as its theme. Italian design continued to be successful. In the early 1980s, while attention was focused mainly on Memphis, a new movement got under way: Neoprimitivism – a formal, metaphorical reclamation of the primitive state.

The Domus Academy

After 15 years of 'new design', ranging from radical architecture to primary design and from Alchimia to Memphis, Andrea Branzi set up the Domus Academy in Milan in 1983. The best designers came together at this school, which functioned as a kind of observation point for international design, a place of confrontation and debate. During its first three years, the Domus Academy ran a course of study looking at major contemporary problems. Its students came from 80 countries; the teachers were masters of Italian design – Mario Bellini, Ettore Sottsass, Rodolfo Bonetto, Achille Castiglioni, Vico Magistretti and Richard Sapper. In a spirit of cultural exchange, the school invited in 'associate professors'. It also organized seminars on technical design and visits to businesses. Research was carried out in collaboration with companies such as Abet Laminati, Fiat, Flos and Kartell. An ambitious project, the Domus Academy took as its basis a theoretical document which summarized the themes it wanted to explore.

Neoprimitivism

Andrea Branzi continued his exploration of new-style craftsmanship. In 1985, he had the idea of bringing together, in a free association, natural elements, birch branches, sections of tree trunks and a standardized base which would form the seat and underframe of a series of chairs. He called the collection *Animali Domestici* (Pets), referring to the paragraph 'Le plus bel animal domestique' in Jean Baudrillard's first book, *Le Système des Objets* (1968). He described the style as 'Neoprimitive'. He also designed a sofa and chairs which were brought out by Zanotta as part of their *Zabro* collection (1985). The influence of Arte Povera, an Italian avant-garde art movement, can be detected in his use of raw materials. The collection explored the domestic condition – in other words, the set of behaviour patterns, affects and psychological values that are to be found within the home. The home was taken as the starting point for every project; it was viewed as a sort of anthropological site. 'We have to look anew at the rituals, myths and magical properties of our surroundings,' Branzi declared. Gaetano Pesce similarly explored radical, powerful themes like those contained in Neoprimitivism.

Italian humour

Denis Santachiara (b.1951) saw in technology a way of filling a home with mischievous spirits. In his work he used a great deal of both imagination and technology. For example, his *Maestrale* lamp (1987) is equipped with a blower that makes a little red flag flutter. With roots in the radical utopia of the 1970s, Alessandro Mendini (b.1931) was one of the founders of Alchimia and one of its most brilliant representatives, as demonstrated by his *Proust* armchair (1978). In the 'Oggeto Banale' (Ordinary Object) exhibition at the Venice Biennale in 1980, he showed – along with Daniela Puppa and Franco Raggi – a series of everyday objects which he had reinterpreted by adding decorative or kitsch elements to them. Together with 30 or so other designers he was involved in making *Il Mobile Infinito* (Infinite Furniture, 1981) – a limitless series of furniture items, a construction set with no end. The Florentine designer Pierangelo Caramia (b.1957) was one of the founders of the Bolidismo movement, which was established in Bologna in July 1986 but proved to be short-lived. It was inspired by the designs of Norman Bel Geddes and the Streamline Style, and laid claim to modernity on account of its using electronics as a means of expression. The Yves Gastou Gallery staged the first Bolidismo exhibition in Paris in 1987. Caramia produced stylistically nostalgic work: for example, his little glass and aluminium table *Arcadia Swing* (1987), manufactured by XO; its underframe is shaped like the Statue of Liberty.

Conceptual design

Furniture became the subject of intellectual speculation. The 1980s saw the emergence of a generation of conceptual designers who thought deeply about the concept and meaning of objects.

Contrary to the principles of corporate design, the new designer set out to be an interpreter. He spoke the language of myths and the senses rather than of slogans and logos. He was interested in the discourse of society, not that of business. The advent of electronics made it possible to design objects that were both useful and attractive. While traditional design sought to provide aesthetics with a scientific basis, 1980s design chose, conversely, to provide scientific fact with an aesthetic basis. The renewal of design involved challenging the idea of pure functionalism: the aim of designers was to go beyond the mere usefulness of an object. Designers from the Memphis group – George James Sowden, Marco Zanini and Michele de Lucchi – joined Ettore Sottsass at Olivetti. They treated industrial objects and electrical domestic appliances in a playful manner. These practitioners of 'new design' rediscovered the benefits of large production runs. The work of Achille Castiglioni (1918–2002) belonged to the tradition of free functionalism. He designed functional products with pure lines: the *Gibigiana* standard lamp, manufactured by Flos (1980); the cruet set manufactured by Alessi (1980); the *Ovio* and *Pario* glasses for Danese, and sanitary ware for Ideal Standard.

Britain in the 1980s

The punk phenomenon produced fashions and imagery that reflected the most anarchic tendencies of youth. It had a profound impact on designers who had recently left art school; these at last put Britain at the international forefront of design. Following the design seminar organized by Margaret Thatcher in 1982, the government trebled the annual grants for promoting creativity in design, and design experts set about redesigning shops throughout the country. Design became a real service industry, and the word 'design' was used as a marketing term. But the market remained relatively limited. Design was not much used by industry, and the scene continued to be very much dominated by one-off items and limited runs.

Freelance designers

Designers gave up collaborating with industry and tried to make and distribute their products themselves. The absence of any economic pressure allowed them greater freedom and creative audacity. The Architectural Association School of Architecture in London produced architect-designers like Nigel Coates (b.1949), who founded the NATO (Narrative Architecture Today) group. In 1987, he launched the *Jazz* and *Metropole* furniture ranges, followed by the *Noah's Ark* collection (1988) in Milan. The architect Zaha Hadid (b.1950) also took an interest in furniture. Her work – for example, *Red Sofa* (1987–8) – brings to mind the lines of the Russian Constructivists of the 1920s.

The underground

In the field of fashion design, there were many young designers who turned the youth style into an international style. In the 1980s, London was home to two mutually antagonistic design trends at the same time.

Interior of the One Off gallery-workshop, Ron Arad, 1981. Stainless steel furniture. Paris, CCI-Georges Pompidou Centre/Kandinsky Library.

The first of these was represented by Julian Powell-Tuck, Danny Lane, Ron Arad and the NATO group. Its most powerful sources of expression were devastation and breakage – the aesthetics of destruction. Inspired by the punk movement as represented by John Richmond and the fashion designer Vivienne Westwood, it featured decrepitude and cracked or broken objects. It was like an archaeological investigation into the first Industrial Revolution. Specialist galleries exhibited the works of these freelance designers. Together with Tom Dixon and Peter Keene, Ron Arad (b.1951) opened the One Off design office in Covent Garden (studio, workshop and showroom). He designed furniture described as high-tech and 'ruinist': works such as *Transformer, Rover, Concrete Stereo* and *Aerial Light.* He also created architectural elements, and exhibited with Tom Dixon, André Dubreuil and Danny Lane. Their works, produced in limited runs, were regarded as usable art. At the 'Nouvelles Tendances' exhibition at the Pompidou Centre in Paris in 1987, Arad brought real

DC02 vacuum cleaner, James Dyson, 1995. One of the last models in the innovative Dyson range of bagless vacuum cleaners. The first model, the *G-Force Cyclonic,* revolutionized household electrical goods in 1983.

life back into his work: in order to abolish design, he invented a large machine – a powerful crusher – for demolishing old objects, and old chairs in particular.

A door would open and a brick equivalent to ten chairs would be automatically ejected: one more brick for the wall of chair bricks. Arad was drawing our attention to the object that refuses to die. Jasper Morrison (b.1959) argued the case for balanced design accompanied by a kind of return to morality. His *Wingnut* chair (1985) and his plywood desk (1988), manufactured by Neotu, are outstanding.

The second trend, which might be described as Neobaroque, was represented by Tom Dixon, André Dubreuil and Mark Brazier-Jones, and was echoed in John Galliano's fashions. The extravagant furniture of Tom Dixon (b.1959) was hand-made. He welded together industrial objects to create sophisticated pieces of metal furniture that were eccentric, baroque objects, and stylistic references. André Dubreuil (b.1951), a French designer living in London, took his inspiration from 18th-century French furniture and the Baroque movement, but also used rough materials such as reinforcing steel.

He exhibited in London and Paris, acquiring an international reputation and receiving many commissions. Mark Brazier-Jones (b. 1956) founded the Creative Salvage group with Tom Dixon. He favoured materials such as cast iron or aluminium, which he gilded. He was inspired by mythology to create such items as the *Nemo* cabinet and the *Arrow*, *Atlantis* and *Whale Tail* chairs. The introduction of art into design was a major preoccupation for all these designers.

A generation of 'industrial designers' who came up with sometimes complex projects emerged in Britain in the early 1980s. Daniel Weil (b.1953) wrapped his radios, clocks and calculators in plastic bags (1981). Abandoning rigid, opaque casings, he refocused on the beauty of the electronic components through the transparency of the bags. Winfried Scheuer came up with a prototype television, a fan and a radiator. Some objects were notable for their pure aerodynamic design, such as the *G-Force* vacuum cleaner by James Dyson (1979); after years of work and research, it was finally manufactured in Japan.

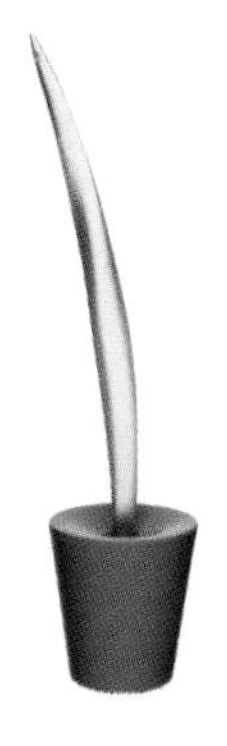

Toothbrush from the *Good Goods* catalogue for La Redoute, Philippe Starck, 1998.

Avant-garde design

Functionalism reflects a mechanized society, a society of turbines and bolts. The 1980s were the age of communication and virtuality. One of the most famous headlines in *The Times* announced, in 1984: 'Mr Computer, Man of the Year'. Ever since the production of the microchip by the Intel Corporation (USA) in 1971, microelectronics had permeated society. The miniaturization of elements allowed designers to reduce the size of objects and play with their external form as much as they wanted. In 1985, the American firm Zenith perfected the touch-sensitive screen. The Sony *Walkman* personal stereo, launched in 1979, was staggeringly successful. It marked a bigger turning point in Japanese design than any purely technological innovation would have done. It denoted a new style, a new attitude, a new way of living. The relationship between form and function was no longer the same. The new technologies of the late 20th century made it possible to create a virtual image of the product and visualize on screen how it would function in reality.

Form was decided digitally. Style was universal, determined by set formulas and colours. The idea of modernity went hand in hand with global recognition of a product.

A virtual society

Three-dimensional CAD (computer-aided design) led to profound changes in the working methods of design studios. Computers became the main tools for developing and revising projects. Digital technology was at variance with traditional methods. The efficiency of CAD software resulted in a reduction in the time needed to do things, and made it possible to offload many humdrum tasks (plans, digital prototype development, imaging, and so on), thereby freeing up time for pure creative design. On the other hand, computerized equipment required in-depth training, and there was a quick turnover of software, which needed to be updated approximately every six months. Within a technological world where information systems were all-powerful, computerized equipment itself had its own aesthetics. The form of a computer did not reveal how its components (microprocessors, memory, miniaturized elements) were put together; there was no visible explanation of how they operated. In fact, the exact opposite was the case: their end purpose was concealed. Form no longer followed function, but had to symbolize it. This technological advance created a historical distance between the objects designed at the end of the 20th century and the tools of the first industrial age, which were often characterized by their functional conformity.

Several personal computers (PCs) were available on the market in the late 1970s, but it took a long time to master the necessary computer language. In 1976, Steve Jobs and his partner Steve Wozniak founded

Biodesign

Frog, proposal for an underwater camera for Canon, Luigi Colani, 1974. The forms of the camera harmonize with those of the human body. An example of biodesign.

'Nature is the starting point': this was the central idea behind the biodynamics of Luigi Colani (b.1928), one of the most controversial of designers. Some people see him as a professional critic or a design 'entertainer', but he scores highly with others, who regard him as a philosopher and a genius. He created products relating to car, motorcycle and aircraft design, but also useful everyday objects such as sports equipment, furniture, bathroom accessories and cameras. The most striking example of the latter is the Canon *T90* camera. Colani produced a design that harmonized with the human body and even seemed to be a part of it. It was awarded the accolade of 'Camera of the Year 1987'. Working for Canon, Colani made full-size mock-ups of reflex cameras such as the *Hypro*, which looks like a shell, and the *Frog*, an amphibian that looks like something out of a science-fiction comic strip. He designed the *MX5* – a round-shaped 'concept car' of the future – for Mazda. Biodesign is curved and organic, and encourages us to think about the physical relationship of the object to the human body. Based on sinuous forms and backed up by ergonomic research, this stylistic language attracted many followers, who designed such products as the Panasonic *Videophone*, the Yamaha *Morpho* motorbike and the Olympus *AZ330* camera (1988). Biodesign also affected camcorders, CD players and radios. The perfecting of these complex forms was made possible by CAD (computer-aided design). Sony invested heavily in this technology, designing its own software, Fresdam, on a Silicon Graphics station, before entering into a research contract with the University of Michigan relating to the optimization of forms.

the company Apple Computer. They challenged IBM and other large companies by producing a range of equipment that was not IBM-compatible. Their aim was to make a PC that was simple to use: the Macintosh. The first models were called *Apple II* (1977) and *Lisa*, and they were already very compact. One of the elements specific to the Apple Mac was the mouse, which supplemented the keyboard. The other novelty was its user-friendliness; it used simple words, graphics and symbols associated with the traditional office. Created at the beginning of the 1980s, this computer was launched in 1984 with a wide-ranging advertising campaign orchestrated by the film-maker Ridley Scott and associating the product with the notion of freedom. In 1982, Harmut Esslinger's agency Frogdesign was chosen to run Apple's official design programme until 1985. After that, other agencies became involved. In 1989, it was Giugiaro Design that designed the first versions of what would become the Powerbook.

In 1990, the Internet became a global network dedicated to non-military research. CERN (Conseil Européen pour la Recherche Nucléaire) was behind the development of the World Wide Web, which allowed the general public to access the Internet.

Corporate design

There was a proliferation of modular office equipment along the lines of Herman Miller's *Action Office*. Economic prosperity triggered renewed interest in 'corporate design'. The companies Vitra, Herman Miller and Steelcase Strafor focused their research and investment on the office

Glass and chrome-plated steel desk from the *Nomos* range, Norman Foster, 1986. Manufactured by Tecno. Saint-Étienne, Musée d'Art Moderne.

Aeron chair, Bill Stumpf and Don Chadwick, 1994. Manufactured by Herman Miller. Houston, Museum of Fine Arts.

environment. Several types of adjustable office chair were produced, aimed at both secretaries and directors.

Emilio Ambasz (b.1943), an Argentinian designer, and Giancarlo Piretti (b.1940) designed the first ergonomic office chair for secretaries: the *Vertebra*, which was awarded the Compasso d'Oro in 1981. The German company Vitra, which opened its design museum (built by Frank O Gehry) in 1989, produced the two upholstered office armchairs *Figura* (1985) and *Persona* by Mario Bellini. In the 1980s, famous architects turned their attention to the technical design of office furniture. The English architect Sir Norman Foster (b.1935) designed the *Nomos* furniture range (1986), made from glass and chrome-plated steel, for the Italian company Tecno. He received the Compasso d'Oro for it in 1987. He put the case for his design as follows: '*Nomos* is not a set of tables, ... it is first and foremost a way of thinking, of creating surface as function ... that is to say, [a space where] human beings can carry out their public and domestic rituals.' Architect-designers had become 'high priests', capable of bestowing meaning on objects. Richard Sapper's elegant *9 to 5* (1987) for Castelli was derived from this new idea of 'ritual' that was now an integral part of designers' language.

The same notion was behind the design of the *Ethospace* system (1986) for Herman Miller by the American designer Bill Stumpf (b.1936). Employees were provided with an open, high-quality space that was flexible enough for both personal and collective use.

A technology of the future: (im)materials

From the 1980s onwards, designers began to concentrate on industrial objects, and this culminated in the exhibition 'Les Immatériaux', organized by the philosopher Jean-François Lyotard at the Pompidou Centre in 1985. He was interested in the dematerialization of objects which could be brought about by new techniques such as fibre optics.

Two new materials appeared in the 1980s that would encourage a return to craft-based work: MDF (medium density fibreboard), which could be used in plank form like wood, and ColorCore, first manufactured by Formica in 1982. In order to publicize its product, Formica commissioned artists and designers to create furniture that would be shown at an exhibition in Pittsburgh in 1983.

The American designer Stanley Tigerman (b.1930) made the most ingenious use of the material.

In the early 1980s, a series of more technical pieces of furniture appeared, such as the *Sinbad* armchair by Vico Magistretti, which was presented by Cassina at the Milan Furniture Exhibition in 1981 and which had a woollen blanket thrown casually over the seat as if over a horse's back, or the *Wink* armchair (1980) by Toshiyuki Kita (manufactured by Cassina), with a completely elastic structure in the headrest area. The *Penelope* chair, designed by Charles Pollock and made by Castelli, was the hit of the Milan Furniture Exhibition of 1982. It consisted of a continuous, 6-metre-long steel strap, onto which a seat of plaited steel mesh had been fixed. In the same spirit, the German manufacturer Vitra concealed ultra-sophisticated technology under shapeless cushions in *High Touch*. The architect Jean Nouvel made systematic use of aluminium. His work marked the transition from mechanical to electronic technology: for example, his *Telescopic Table* (1987), a prototype funded by the VIA.

The appearance of halogen in domestic lighting demonstrated how industrial technology could be adapted to the home environment.

Unfortunately, it was hard for young French designers to gain access to new technologies and materials. Kevlar, carbon, honeycomb aluminium, liquid crystal and so on were not affordable for use in domestic interiors, and even less so in the field of furniture. Nevertheless, there was always bound to be some practical spin-off from leading-edge research. Space research was still in its infancy, but it would throw up more and more new materials and technologies for new and different lifestyles. In December 1985, Jerry Ross and Sherwood Spring assembled aluminium structures that were put aboard the 23rd Atlantis space shuttle. Living on the Moon or in a space station no longer belonged to the realm of science-fiction. Leading-edge technology, and astronautical technology in particular, was exploited by some designers: Alberto Meda used carbon-fibre resin for his *Light Light* chair (1987, made by Alias), which weighed only one kilogram. However, this was still a very expensive material.

A return to craftsmanship

All the same, the 1980s marked a move away from the industrial approach by creative designers. For the most part they designed furniture they could make themselves, using carved wood or folded metal. This self-sufficiency gave them much more freedom. A return to craftsmanship is one of the characteristics of the period, along with pluralism and a tendency for designers to work on a freelance basis.

In the USA, Wendell Castle (b.1932) made decorative elements and items of furniture from sawn wood, put together by hand.

His furniture is imposing in its form, and the details are full of sensuality: the Egyptian desk (1982) made from maple and ebony; the *Dr Caligari* pedestal clock (1984), made from cherry wood decorated with ebony; and *Ghost* (1985), an old clock clad in mahogany. Also in the USA, George Nakashima (1905–90) designed works that made reference to 19th-century Shaker furniture and Japanese handicrafts. These pieces, the opposite of mass-produced items, were extremely expensive. This return to 'work well done' and objects that could be handed down from generation to generation showed that many consumers were tired of disposable, expendable industrial products. In 1976, John Makepeace (b.1939) founded a school of woodcraft in Britain, opening one of the best-known woodcarving workshops in the world. He thus renewed links with a tradition of craftsmanship in the spirit of the Arts and Crafts movement.

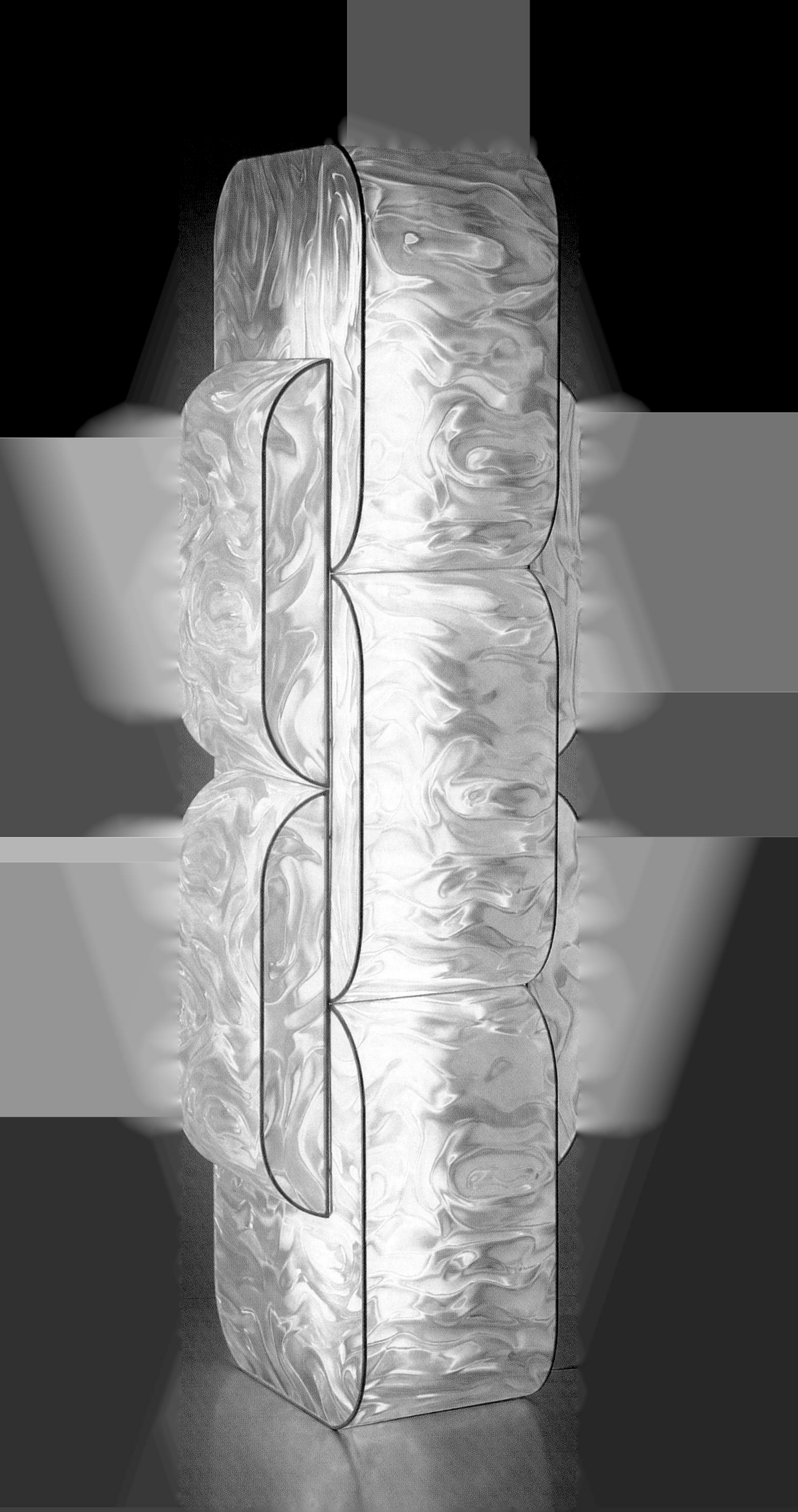

1990-2005: a virtual era

- The cult of the object
- Design and brand strategy
- Sustainable development and the environment
- A new generation
- Interactive design
- Design-fiction

'What do we do after the orgy?' (Jean Baudrillard, 1990). The effects of the First Gulf War on the economy, the worldwide spread of Aids and a whole series of ecological problems threatened the viability and long-term survival of the planet. It became essential to produce objects that protected the environment at every stage of their existence. Consideration was given to the materials deployed, to the energy used in producing, operating and recycling goods, to ergonomics, to aesthetics and to the recycling of materials. In addition, the 1990s were characterized by a policy of global industrial regrouping. Geographical frontiers were replaced by interest groups. In France, design was decentralized through various bodies and activities: the CIRVA (Centre International de Recherche sur le Verre), based in Marseilles; the CRAFT (Centre de Recherches sur les Arts du Feu and de la Terre), based in Limoges; the Biennale at Saint-Étienne (1998); and initiatives organized by the Villa Noailles in Hyères. As the 20th century drew to a close – a time when creative energy was being wasted in virtual digressions – design rediscovered its roots. Several retrospectives were organized: 'Design, Miroir du Siècle' at the Grand Palais in Paris in 1993; 'Roger Tallon' at the CCI in Paris in 1993; 'Design Français: l'Art du Mobilier 1986–96', a homage to rational design, at the cultural centre in Boulogne-Billancourt (1996); a homage to Charlotte Perriand at the Design Museum in London in 1996; and 'Europa, Europa' in Bonn (1994). Design was reclaiming its history and reinterpreting legendary objects or brands. The Laguiole knife was revisited by Philippe Starck, Louis Vuitton brought out a folding table made of natural leather by Christian Liaigre, and Hermès produced a new line of luggage by François Azambourg. Ever since the 1980s, designers had regarded the luxury industries as new commercial partners – while still remaining faithful to their mission of 'serving everyone'.

The cult of the object

Nowadays, the design discourse is carried on by non-designers: the media, marketing and communication experts, sociologists, philosophers and semiologists. Marketing and communication have introduced concepts of emotion, sensuality and enjoyment into the discourse about the object – concepts which are far removed from technological and functionalist considerations. Marketing studies focus on psychological values rather than utilitarian ones. Design is no longer called into question, because it is fully integrated into the logic of marketing and image. Designers are required to come up with pleasing, attractive products. Design now concentrates on the immaterial. A fundamental notion called a 'look' gives objects their power of seduction. This comes under the heading of styling, which is an aesthetic exercise, and hence an artistic one.

Previous page: *Chain* lamp, assemblage of five elements with a hologram effect, Tom Dixon, 2003. Made by Tronconi.

Art and design

At the end of the 20th century, the relationship between art and design was reversed. The time when artistic avant-gardes influenced design was past. It was the turn of design to become the model for art. In the 1990s, installation art was again topical. Franz West's work *Auditorium* (2000), for example – an installation of sofas and rugs – questioned the function of art. Many works of art could have been mistaken for living spaces given a critical treatment by artists such as Jorge Pardo, Tobias Rehberger and Philippe Parreno. The preoccupations of the designer and those of the artist have several things in common: the intellectual working out of ideas, the positioning of one's work in the contemporary creative field, the use of computers, and the establishment of a communications, marketing and production strategy. Just like any artist carrying out a commission, the designer creates a product that will meet the wishes of the person it is intended for. He or she designs the object in a clearly identifiable and (hopefully) distinctive manner, giving it that extra little bit of 'soul'.

Artists have also engaged in the field of design: for instance, Damien Hirst with environments such as his bar-restaurant, *The Pharmacy* (1997). Ron Arad was commissioned by the magazine *Domus* to design a monumental totem for the alternative exhibition at the Milan Furniture Exhibition in 1997. It was for this too that he designed his *Tom Vac* seat, for which he had to have a mould made using aeronautical technology. The Dutch group Droog Design, founded in 1993, came up with a piece of furniture that was functional, in spite of its artistic claim to be an ephemeral, evolving structure: the chest of drawers *You Can't Lay Down Your Memories* (1996) by Tejo Remy.

Craftsmanship and limited runs

Olivier Gagnère was the first person to move in the direction of high-quality craftsmanship. He exhibited works in glass, made at Murano, at the Galerie Maeght (1989), and then designed a collection of ceramic pieces produced by Bernardaud for the Café Marly at the Louvre.

The CIRVA (Centre International de Recherche sur le Verre), directed by Françoise Guichon, opened in Marseilles in 1986. This unique institution, neither school nor gallery, has 1,400 square metres of exhibition and workshop space, as well as materials and a permanent team of engineers and technicians available to artists and designers of all nationalities. The CIRVA also serves as an information centre on the techniques of glassmaking. Many designers – including Gaetano Pesce, Marc Camille Chaimowicz, Erik Dietman, Giuseppe Penone, Marie-Christine Dorner, Sylvain Dubuisson, Borek Sipek, Élisabeth Garouste, Mattia Bonetti, Martin Szekely and Pascal Mourgue – have made pieces there, often one-off items. Exhibitions – 'Trente Vases pour le CIRVA' (1989) and 'Trente Créateurs Internationaux' (1993), staged by Jasper Morrison at the Musée du Luxembourg in Paris – have brought ceramics to a wider audience.

The Daum crystal factory produced designs created by Hilton Mac Connico, Philippe Starck (1988), André Dubreuil (1991) and Eric Schmitt (1995), while the Saint-Louis crystal factory used the designer Jean-Baptiste Sibertin-Blanc.

The CRAFT (Centre de Recherches sur les Arts du Feu et de la Terre) was set up in Limoges in 1993 on the initiative of the French Ministry of Culture at the same time as a new school of decorative arts was opened in the town. The director of the centre, Nestor Perkal, had the task of increasing artists' and designers' awareness of the artistic potential of ceramics. In 1993, the centre produced exclusive pieces by the designers Claude Courtecuisse, Daniel Dezeuze, Olivier Gagnère, Daniel Nadaud and Nestor Perkal. At the Saint-Étienne Biennale in 1998, the CRAFT showed pieces by Martin Szekely, Bécheau-Bourgeois and Éric Jourdan.

The Ministry of Culture and the town of Vallauris organized a joint project called 'Designers à Vallauris' from 1998 to 2002. The ceramic workshops in Vallauris encouraged young designers accustomed to industrial methods to explore the techniques and skills of the craft approach. Many designers took the opportunity to participate in this unique experiment: Martin Szekely (floral bricks turned in raw clay), Roger Tallon (adaptable crockery for taking on holiday), Olivier Gagnère (vases), Ronan Bouroullec, François Bauchet, Pierre Charpin and Jasper Morrison (objects with clean lines), the Radi Designers, Frédéric Ruyant and Patrick Jouin. The artists reserved four numbered copies of their works for the FNAC (Fonds National d'Art Contemporain), and the remainder of the short runs were intended for sale. This passion for crafts was confirmed by the opening of the Viaduc des Arts craft shops and workshops on the Avenue Daumesnil in Paris (12th arrondissement).

'Unmissable' events

The association Designer's Days has the job of promoting design and bringing together all the various players needed to turn Paris into the capital of design. The major events in Paris are the Paris Furniture Exhibition and the International Design Exhibition 'Now'. Another event, entitled 'Le Parcours', began in 2003. Seventeen trade exhibitions relating to fashion and the home are grouped together under the banner 'Paris, Capitale de la Création', which organizes design, fashion and art events twice a year. France established an International Design Biennale in Saint-Étienne, a town with modern industries and factories, such as the Manufrance arms and bicycle factory. The very first biennale was held in 1998 at the Musée d'Art Moderne, which has a fine collection of 20th-century objects.

This international biennale aims to highlight the many fields

Carafes made at Vallauris in the turnery of Claude Aiello and in Martial Quéré's workshop, Ronan Bouroullec, 1999. Paris, Galerie Peyroulet.

explored by design. It presents a wide range of contemporary objects which are emblematic of consumerism, derived from new materials and using innovative technologies. The exhibits come from some 50 countries spread across all five continents. Design is viewed as a global phenomenon. 'By displaying objects in their rich and ever-changing diversity, the Saint-Étienne Biennale will facilitate an 'archaeological' study of the present time and highlight designers' philosophical beliefs, based on the utopias and the anxieties of our civilization,' Jacques Bonnaval wrote in the preface to the catalogue of the 1998 Biennale. In spite of being held in a period dominated by economics, this event shows that design is not always thought of in consumerist terms; it remains anchored in the complexity of its cultural foundations, and is capable of reflecting concerns about the future of the planet and our growing awareness of its fragility.

The annual Milan Furniture Exhibition is an unmissable design event. Sponsored by the magazine *Domus*, this mecca of showmanship encourages an artistic approach to design. In 2001, the designers Tom Dixon, Michael Young, Karim Rashid, Ron Arad and Michele De Lucchi engaged in a live experiment: making items using multi-coloured plastic filaments. New fairs are appearing on the scene, such as the Foire de Courtrai in Belgium. Design is a living entity, and its dynamics can arouse great enthusiasm. Design courses have proliferated throughout the world.

Concept stores and galleries

Design attracts a great deal of media attention, which promotes its spread as a fashionable phenomenon in capital cities and those places where design is 'in'. The Conran Shop (1992) is a temple of designer objects and furniture. The Japanese shop Muji offers high-quality unbranded goods. The shop Sentou makes design classics by designers such as Roger Tallon or Isamu Noguchi, while also developing the work of a range of young designers. In Paris, the shop-cum-gallery Colette (1997) is the arbiter of new, smart international 'good taste'. Also worthy of mention are Tsé-Tsé, the boutique at the Pompidou Centre and the VIA with its private views under the auspices of Gérard Laizé. Design captures the interest of contemporary art lovers at the Galerie Kreo, located in the Louise Association galleries in Paris. Kreo, both a gallery and a maker of design objects, exhibits many projects by contemporary designers. The gallery has exclusive rights over limited-edition pieces by Martin Szekely, Ronan and Erwan Bouroullec and Marc Newson, as well as being the sole distributor for works by Ron Arad.

In their backyard at La Garenne-Colombe, the two master craftsmen Bruno Domeau and Philippe Pérès are responsible for making and distributing works by designers such as Christophe Pillet, who was the first

Very Nice chair made of birch plywood with a leather seat pad, François Azambourg, 2004. Made by Domeau & Pérès.

to entrust his designs to these skilled cabinet-makers. Matali Crasset, Ronan and Erwan Bouroullec, Andrée Putman, Alexandre de Betak, Jérôme Olivet, Élodie Descoubes, Laurent Nicolas and others have also put their trust in the quality of the two men's work. A recent project of theirs is the *Very Nice* table and chairs (2004) by the designer François Azambourg. For this complex, ambiguous work Azambourg produced drawings and calculations to effect a subtle alchemy: a bundle of birch plywood which, once assembled, becomes an architecture of the void, a trap for light.

Pierre Staudenmeyer was the first person to get involved in this trend towards short production runs in the early 1980s. He was a product-maker who also organized exhibitions. Recognized as an astute observer of the period, he published many books. He closed his gallery Neotu, and opened the gallery Mouvements Modernes on the Rue Jean-Jacques-Rousseau. Several other Parisian galleries – for example, Chez Valentin and De Di By – were also very interested in contemporary design.

Concept stores sprang up. These aimed to surprise a public jaded by overabundance and rekindle its enthusiasm. Design emerged from museums and galleries and took over the boutiques. In New York, the boutique at the Museum of Modern Art and the boutiques Murray Moss and Totem were virtually like design schools. In London, Space (fitted out by Tom Dixon), Oggetti, the Design Centre boutique, Coexistence and the Conran Shop were prestigious showcases for contemporary design. In Tokyo, there were the dynamic production facility Idée and the Axis centre, which held exhibitions. The city hosted 'Design Week' one year, providing an opportunity to discover the vitality of E&Y, Trico and the Harajuku Gallery – all young producers and distributors of design objects. Influenced by European creative designers, the younger generation devoured international design publications.

In the USA, Evans and Wong introduced a new sales concept. Christopher Evans and Victor Wong published a catalogue which they distributed to private clients, cultural institutions, businesses, and journalists in Europe and the USA. It was in effect a reference manual of exclusive, very contemporary works by young designers. Evans and Wong operated in a particular way, holding minimum stocks and paying their suppliers only when the goods were sold. The trial issue of their *Approximations* catalogue was published in 1995 (2,000 copies), as were the first issues of *Brainstorm* (3,000 copies) and *Copyright* (10,000 copies). Their novel approach aroused people's curiosity. They also set up a website listing their products. The French designers of Appartement D came up with a similar idea: 'A non-profit-making association, 30sq m, four founding members and ten other members.'

They mounted a fringe exhibition at the Furniture Exhibition of 1998. Their slogan was: 'If you're a buyer, buy! If you're a designer, call us!' They concentrated on production and sales. They published a mail-order catalogue, and offered young designers the chance to make prototypes of their designs.

Food and design

The latest 'fashion attitude' is an interest in food, a trend started by the Costes brothers when they opened the Café Costes in the 1980s. In addition to new tastes or new attitudes towards food, food-centred design encouraged the development of new eating places. Philippe Starck created a new concept with the restaurants Bon (on the Rue de la Pompe in Paris) and Bon 2 (on the Rue du Quatre-Septembre), making Jean-Marie Amat his consultant chef. Restaurants of this type proliferated: Food Unlimited in the Beaubourg district (furniture by Christophe Pillet); the Georges at the Pompidou Centre (interior by Jacob and MacFarlane); Lo Sushi at the Pont Neuf, with a large counter made of Corian and designed by Andrée Putman; Terence Conran's restaurant; the Versace restaurant-boutique; the Café Beaubourg; the Top Cloud restaurant in Hong Kong, designed by Mathilde Bretillot and Frédérique Valette and serving 'fusion food'. At the restaurant R'aliment on the Rue Charlot – the headquarters of the designer Philippe Di Meo – the food is organic and colourful. Patrick Jouin, working for Alain Ducasse, designed the bread rolls for a new type of sandwich bar (2000), and fitted out his restaurant Mix on 58th Street in New York. Claudio Colucci fitted out the Delicabar at Bon Marché. The concept for this smart snack bar was developed by the young pastry chef Sébastien Gaudard. Further evidence of the pleasure to be derived from cooking was provided by the recipe book produced by the magazine *Case da Abitare* in 2003. This contains 22 recipes by designers, including Tord Boontje's 'Blackbird Pie', Matali Crasset's 'Op Cake', Massimiliano Fuksas's 'Boom!!!' and Massimo Iosa Ghini's 'Torta a Due Piani'.

Design and brand strategy

As both a fashion phenomenon and a social phenomenon, design has become a favourite vehicle for brand strategy. Publicity and marketing have very specific aims, and to this end the designer is required to be an interpreter and spokesperson for the business.

Designer everything

In 1993, the new director of the Philips design studio, the Italian designer Stefano Marzano, announced that he intended to promote the Philips brand identity through design. He created the Visions of the Future collection, including the *Daisy* mini-vacuum cleaner, the *Billy* mixer and the *Bob* kettle (1995–6).

These followed in the tradition of the 'dream products' designed by

Alessandro Mendini and manufactured by Officina Alessi. Other design companies pursued various ideas for exploring the object. Moulinex organized the competition 'Génération Design' (1990) for European students. The Italian company Zanotta encouraged design creativity with the help of Roberto Pezzetta, who designed the *Oz* refrigerator (1996) with its sensuous curves. Marc Berthier and Elium Studio worked on ranges for Rowenta, Lexon and Magis.

No sector was overlooked. Even the French electricity company EDF launched a competition to find designs for their high-tension pylons (1994). Many teams entered the fray, including Giugiaro Design, Starck-Méda, RSCG-Tallon and Wilmotte-Technip. Designs by Marc Mimram's French team and the English team Ritchie-RFR-Gustafson were selected. The French company Ricard asked Garouste and Bonetti to design a carafe (1995) and a bottle (2000), while the Radi Designers were approached to create the dispenser (1999). The company Jean-Claude Decaux began innovative research into street furniture: some designs by Philippe Starck, Norman Foster and Martin Szekely were put into production. In 1996, Starck left Thomson's Tim Thom Studio which he had set up in 1993 by merging a team of young designers with the existing team. The studio carried on, coordinated by Matali Crasset, Éric Jourdan and Patrick Jouin. They created 'dream products' such as the *Cub* and *Vertigo* LCD projectors, the *Rock & Rock* hi-fi system, the *Alo* digital telephone, and the *O Clock* radio alarm clock. Further collections were developed by the studio. The 'Line', 'Partenaires' and 'Gamme' collections were also produced by the Tim Thom Studio, meeting fixed specifications and constraints relating to manufacturing, production runs and distribution.

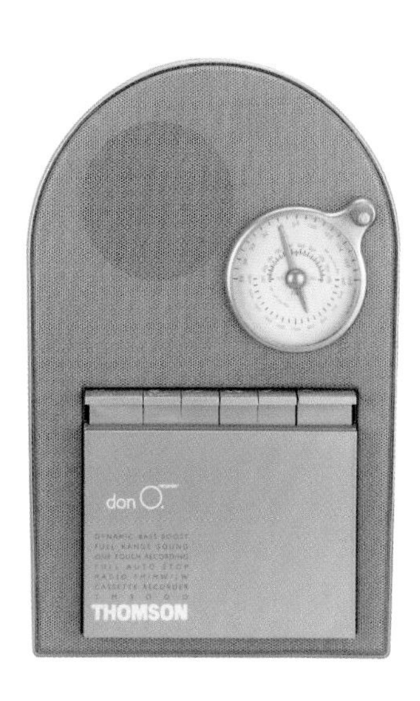

Don'O mono radio cassette player, 'Line' collection, Matali Crasset, 1993. Tim Thom Studio for Thomson Multimedia.

Concept cars

Renault launched the concept car *Scenic*, and Patrick Le Quément designed the Renault *Twingo* (1992). From 1995 onwards, the aesthetics of the 'new edge' style – based on broken lines – took over from biodesign. Claude Lobo, a designer with Ford, was the first person to put this style into practice, along with the Italians who designed the Lancia *Ypsilon*. Sales picked up, as all models began to be inspired by the *Twingo*, including the Mercedes *Smart*. The *Smart*, a tiny runabout, was the outcome of a collaboration between Mercedes and Swatch. Its overall dimensions are small due to built-in bumpers. Safety is a high priority, with the passenger compartment forming a 'Tridon' safety cell, which is visible from the outside. The interchangeable bodywork elements can be recycled. The *Smart* has a playful appearance, as if it had just driven out of a cartoon. In 1999, the Australian designer Marc Newson created the *021C* concept car for Ford.

Sport and the cult of the body

Sports goods are both high-performance specialized items and expendable consumer products that respond to the desire to be fashionable and to conform to a certain 'look' (as demonstrated by the fashion for trainers and tracksuits). Music, advertising and the star status of athletes are effective means of influencing the public. Innovation in sport is synonymous with high technology. High-performance materials are being continuously developed. In sailing, for example, such materials are needed for competition-standard catamarans, trimarans and monohulls. For the Barcelona Olympic Games in 1992, the designer Mike Burrows came up with a racing bike made from resin reinforced with carbon fibre and strengthened by titanium inserts. The sports-equipment market (surfing, snowblading, rollerblading, and so on) reflects a new culture based on freedom, individualism and the quest for pleasure and sensation.

Sports shoes flooded the international market during the 1980s. Although they became the emblems of urban tribes (rappers, skateboarders, and so on), they could just as well be used for walking around town. Adidas launched a range of sports shoes under the slogan 'Feet You Wear' (1997), working with the designer John Earle, who created the *Training Mercury* shoe (1998) with its torsion system. The shapes and colours of the range were amazingly kitsch and playful. The result was a high-grade technological product adapted to the contours of the foot. Adidas's major American competitors, Nike and Reebok, came up with similar products.

The Nike company was founded by Phillip Buck in 1962. He started off by distributing sports shoes imported from Japan. Together with Bill Bowermann he developed a sole injected with latex, which he incorporated into the *Moon Shoe*, intended as a running shoe. In 1979, the

Lotus competition bike, Mike Burrows, 1991.

company developed the air-cushion system, which softens the impact of the foot on the ground; then it perfected the sprung system with *Air Max*, which provides added bounce with maximum comfort.

Nike stimulated sales by launching large-scale publicity campaigns, accompanied by special music and the slogan, 'Just Do It.' The logo had a very powerful effect. The brand worked out a totally innovative distribution policy based on Nike Towns – huge shopping centres and cult consumer destinations. Nike produced iconic products. In 2004, the company approached 25 Japanese artists and asked them to interpret the *White Dunk* shoe in any way they wanted. The resulting exhibition was held in Paris at the Palais de Tokyo, within the Centre de Création Contemporaine, thus lending the product the status of a work of art.

Materials and surfaces

Once it had been integrated into the industrial fabric, design was increasingly affected by new developments in raw materials, as evidenced by the exhibition 'La Matière de l'Invention' at the Pompidou Centre (1989), which anticipated that the future of design would be dependent on new materials. Companies were formed which specialized in creating inventories of state-of-the-art materials. The first of these, Material Connexion, which was formed in New York in 1997, established categories of materials: ceramics, glass, polymers, carbon-based products, cement-based products, natural products and by-products. A library of 3,000 samples complete with technical data was made available to researchers, designers and architects. The agency was slanted towards sustainable development and also provided technical data sheets. Innovathèque, an offshoot of the CTBA and the VIA, was set up in Europe, based on the same principles. Other private initiatives followed.

At the 2001 Furniture Exhibition in Paris, UNIFA (Union Nationale des Industries de l'Ameublement) presented a survey of French creativity in the 20th century, considered from the point of view of raw materials and core materials. The list included pressed steel, solid glass, composite (Martin Szekely's storage cupboard, made by Kreo, 1998), fibreglass (the *Gosthome* armchair by Jean-Marie Massaud, 2000), polycarbonate (the *La Marie* chair by Philippe Starck, Kartell, 1999) and high-impact polystyrene (Ronan and Erwan Bouroullec's shelving in homage to Charlotte Perriand, Galerie Neotu, 1997). In the 'design lab' section of the 2003 Furniture Exhibition, the designer Christian Ghion presented 46 projects carried out in collaboration with businesses that had made new technological materials and resources available to designers. The Italian designer Alberto Meda created an ultralight chair called *Light-Light* (1989), manufactured by Alias.

Such materials relied on leading-edge technologies. Moulding techniques had advanced, and plastic in liquid form was now injected into a turning mould. The company Kartell used this process for the *Bubble Club* chair by Philippe Starck (2000), made from coloured polyethylene. With Jasper Morrison's *Air* chair (2000), made from polypropylene, Magis injected air into the legs, making it possible to do without rungs. For his *Horizontal Chair* (1997), Jean-Marie Massaud used a technopolymer covered with a protein-based self-healing skin which had been perfected in Japan. Titanium, which is reasonably affordable, is used for spectacle frames and some sports equipment. Aluminium has become commonplace. Corian, a new material now used in the kitchen and for bathroom fittings, is a good substitute for ceramic basins. The Hungarian designer Jiri Pecl used it for *Corian Kitchen* (2002), which was exhibited at the International Design Biennale in Saint-Étienne. Arik Levy and Pippo Lionni, working with L Design, have exploited another new material: Alukobon.

Sustainable development and the environment

Consideration started to be given to the dangers of unrestricted economic growth, to utopias past and to dreams of a better society. The preservation of the environment and basic human values and the notion of sustainable development became ever more prominent concerns. Sustainable development describes growth that is capable of meeting 'the needs of the present without compromising the ability of future generations to meet their own needs' (definition of the World Commission on the Environment and Development, 1987). Legislation moved in the direction of making companies responsible for the objects

Cupboard in sheet form, made from Alukobon which is cut out and folded, Martin Szekely, 1998. No screws or bolts. Manufactured by Kreo.

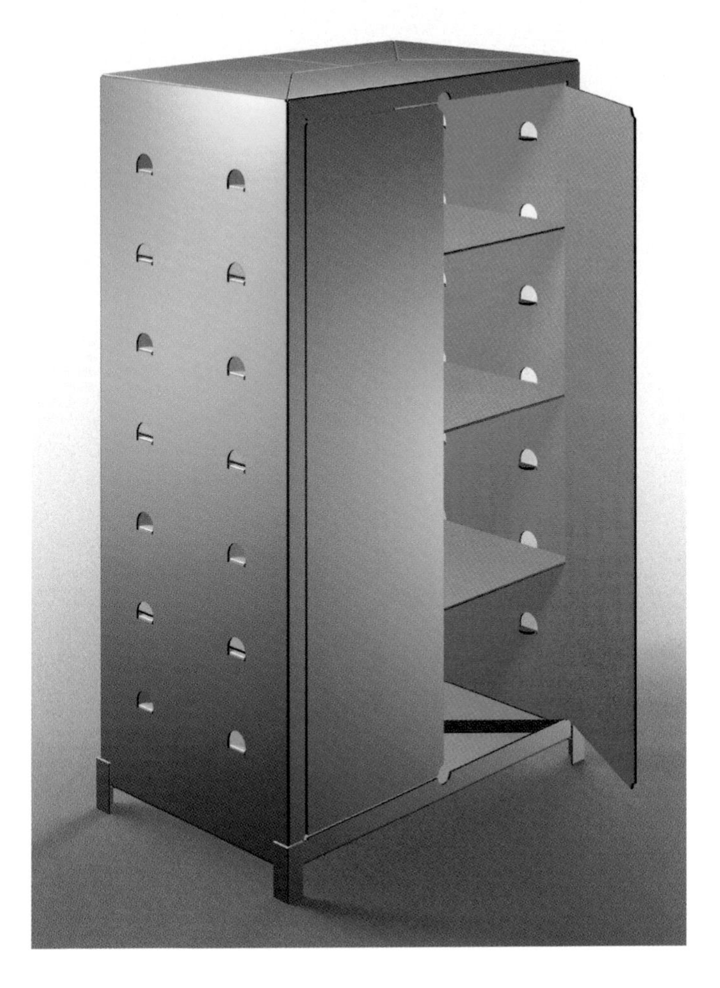

they manufactured, insisting that products should be traceable. Designers played a key role in this policy by contributing their knowledge about how industry operates and about new patterns of consumption. They were in favour of easy accessibility to products, and supported repairing products rather than throwing them away as soon as they stopped working.

An environmental approach could be incorporated into the brief for the design of products. The American concept of 'design for disassembly' was born: this considers, from the very start, how a product is to be dismantled and subsequently recycled.

Good Goods

Philippe Starck has been very much in demand as a designer since the 1990s. 'It is essential to be doing,' he claims. 'I am a citizen who lives, reacts, acts.' He presented more than 200 objects in La Redoute's *Good Goods* catalogue. 'Buying is not the main thing, the interesting thing is to read between the lines,' he said. The catalogue offered 'non-objects for non-consumers ... Honest, responsible objects, respectful of the individual. Not necessarily beautiful objects, but good objects.' This period

Cover of the *Good Goods* catalogue, created for La Redoute, Philippe Starck, 1998. A selection of over 200 objects.

wanted to be a virtuous one. Buyers were looking for basic values, quality and discrimination. Philippe Starck offered products free from hype. He relied on a team of young designers, the best of the 'Net generation': Frédérique Valette, Kristian Gavoille, Bruno Borrione and Thierry Gaugain.

In 1993, Starck took charge of the design department of Thomson Consumer Electronics, which became Thomson Multimedia (a merger of Thomson, Saba and Telefunken) immediately afterwards. For Saba, he designed *Jim Nature*, the first television set made of wood. As he put it, it was essential to 'sell less but sell better' to 'citizen consumers'. He drew up a code of best practice. The small *Bo* chair, manufactured by Driade and made from polypropylene without a single aluminium part, is 100% recyclable.

Ecology and the environment

Many companies actively supported environmental protection. In the USA, the design company Saprophyte devoted itself to research and development in the area of ecologically sound industrial processes.

The exhibition 'Re-f-use: Design Durable/Sustainable Design' was staged at the International Design Biennale in Saint-Étienne in 2002. It was organized by the Dutch exhibition curator Natasha Drabbe, who no longer looked at design solely in terms of aesthetics and fitness for purpose, but also in terms of environmental impact: 'The avoidance or reduction of waste by means of designing sustainable products is our main objective, but the waste produced in spite of these efforts can be put to very good use.' The exhibition consisted of a selection of 150 products from 17 countries; as well as demonstrating sustainable design, they had a strong aesthetic appeal. There was a wide range of materials: bamboo and acrylic, cardboard, paper and plastic recycled into building materials, natural packaging materials, boxes made from citrus-fruit skins, and so on. The Biennale also put forward many other ideas. The O-France agency, in collaboration with the WWF and Victoires Éditions, published the book *Développement Durable au Quotidien*. The same agency has also advised Monoprix on its approach to sustainable development since 1996, and produced a handbook, *Emballage et Environnement* for LVMH (Louis Vuitton Moët Hennessy).

The Swedish agency Kinnarps has been championing environmental issues since 1972. Its slogan is: produce furniture, not waste! Ronan and Erwan Bouroullec designed *Treilles*, a natural partition holding ceramic plant containers, and Fontaine (2003), a ceramic fountain varnished on the inside (made by Teracrea) which supplies water to plant pots. Séverine Szymanski obtained a project grant from the VIA and presented a 'plant partition', *Brike*, at the 2004 Furniture Exhibition in Paris. It is a partition made of clay tiles stacked on a movable plant tub. A system of tanks and tubes ensures that the plants are watered by capillary action.

Luxlab installation, Jean-Marie Massaud, Thierry Gaugain, Patrick Jouin, 1999, prototype (230 x 150cm), made by Via. The designers wanted to create a metaphor for luxury in which the only determining value was a sense of well-being. This extraordinary garden produces three pleasurable sensations: stretching out on the grass while looking at the water and gazing into the fire, thus satisfying primitive basic urges. This is garden art of the third millennium.

Garden art

Patrick Nadeau and Vincent Dupont-Rougier thought up the first 'garden as a piece of furniture', an adaptable, transportable garden-cum-greenhouse-cum-terrace that opens up like a flower. It was presented at the Eighth Garden Festival at Chaumont-sur-Loire, and was awarded the critics' Grand Prix at the 2000 Furniture Exhibition.

Nadeau and Dupont-Rougier made use of hydroponic culture techniques in this domestic design project. Great importance was attached to the design of the tools, so that this type of gardening could be done for pleasure in the home. Technical components such as the watering system were made in noble materials: teak, stainless steel and ceramics. When there are no plants in it, the garden looks like a large closed box, standing 90cm above the ground.

Jean-Marie Massaud, Thierry Gaugain and Patrick Jouin came up with Luxlab, a utopian evocation of a lost paradise, or a return to nature: earth (a floor that can change shape), water (a liquid table) and fire (a meta-hearth). The idea is to rediscover sensations and embrace the 'physical and spiritual pleasure that the elements generate'.

The *iBook*, a portable tangerine-coloured microcomputer, Jonathan Ive, 1999. Made by Apple. A cult object.

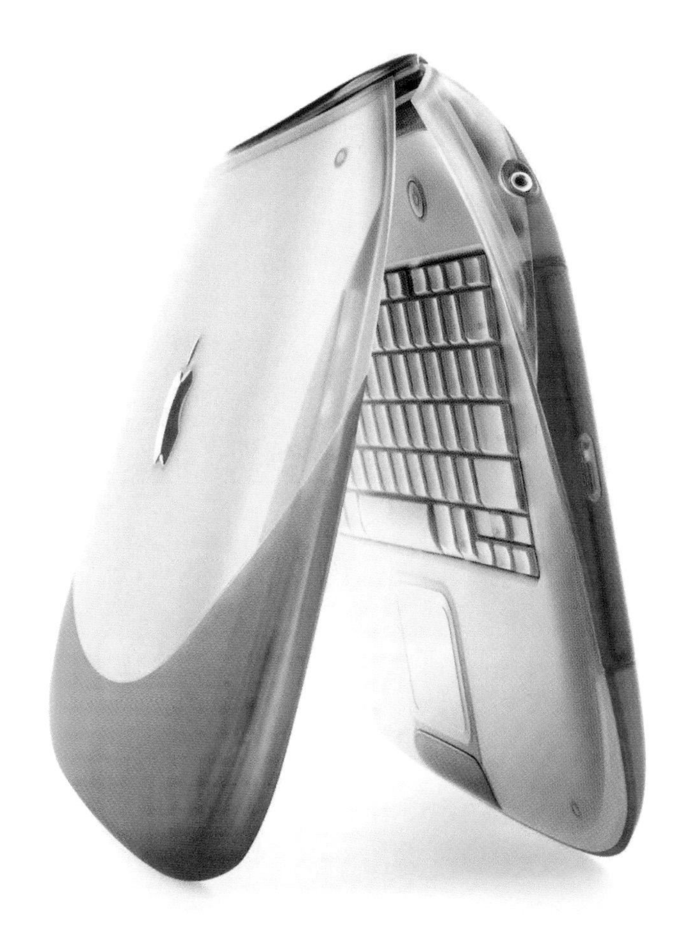

A new generation

European design has remained responsive to changes in society. The notion of mobility has become crucially important, whether it is physical movement in space or virtual movement on the Web. Portable computers have become the central focus of the modern home, supplanting unwieldy desktop computers. Flat-screen TVs and light, folding furniture have also come into their own. Consumers take greater account of wear and tear. They read labels, and pay attention to ergonomic factors. Now much more aware, they refuse to be treated as mere purchasers. A new generation of designers is developing a discreet, versatile aesthetic form aimed at this new type of consumer.

'Low design'

The aesthetics of 'low design' are the province of French designers such as Christophe Pillet, Delo Lindo (Fabien Cagani and Laurent Matras), Éric Jourdan and Martin Szekely and the British designer Jasper Morrison. The book *Modernité et Modestie* (1994, published by the VIA) could serve as a manifesto for the movement. Its representatives include Pierre

Love Seats, cinema seats for two people, made from steel, plywood and moulded foam, Martin Szekely, 2002. MK2/Kreo.

Charpin with his *Slice Chair* (1998, made by Kreo) and Martin Szekely with his Alukobon cupboard (1999), which is assembled by folding and dispenses with screws and bolts. This piece of furniture ushered in a new method of working and constituted a radical design manifesto in itself. It is characterized by its economy of means and exclusive materials, and is shaped by a single gesture, that of folding. Szekely has explored the qualities of many materials: Corian for his *d.l.* desks, boardroom tables, coffee tables and TV tables; birch plywood and cork for the *Cork* chairs (2000); honeycomb aluminium covered with Corian for the *Slim* table (1999) and the *s.l.* table (2003). He has also used plywood and anodized aluminium. Szekely creates designs using basic forms, as can be seen in *Échelles*, a collection of six shelving units (2002). The Galerie Kreo acts as agent for all his designs. He produced *Corolle* (1998), an electricity pylon for high-tension lines made from glue-laminated pine; the Perrier bottle (1996); the floral brick for Vallauris (1998); the *Reine de Saba* (necklace and bracelet, 1999) in silver for Hermès; enamelled porcelain boxes for the Itebos collection, in collaboration with the CRAFT in Limoges; and *Reflet* (2001), solid-silver platters for Christofle. He was commissioned to design the interior of the MK2 Bibliothèque cinema in Paris and the plant beds outside the cinema, and he even came up with *Love Seats* (2003), cinema seats for two people.

A new generation of designers emerged, confirming the strength of French design in certain areas. Evidence of this strength has been on display since 1998 at the Furniture Exhibitions in Milan, Saint-Étienne and Paris. Christophe Pillet (b.1959) produces designs with fluid but restrained lines, as exemplified by the *People* sofa (2004, made by Artelano). He designed a major collection of sofas made by Domeau and

Pérès – *Lobby Sofa, Hyper Play, Nath's Sofa, Video Lounge* – and a system of sofa seating for the Renault showroom in Paris. Jean-Marie Massaud was awarded a Carte Blanche bursary from the VIA in 1995 and produced the *Owen* armchair. He then designed a chair in kit form, the *O'Azar*, manufactured by Magis, and the *Horizontal Chair* (1999), made by E&Y.

He also designed the *Right Stuff* sofa, made by Domeau and Pérès, and a table with an optical effect on its two-coloured leg support (2004, made by FR66).

Often working with his brother Erwan, Ronan Bouroullec is a very private person. All the same, he has made the front covers of design magazines. An advocate of 'bio' design, he was awarded a Carte Blanche bursary by the VIA, which enabled him to carry out a project for an adaptable kitchen (1998). He designed the *Soliflore* vase and a coffee cup made by Evans and Wong. The *Spring* armchair, the *Hole* metal chair and the *Lit Clos* are manufactured by Cappellini. The *Brick Design* storage modules, made from polystyrene for the Galerie Kreo, were subsequently produced in painted wood by Cappellini. The two brothers designed the interior of Issey Miyake's boutique A-Poc (A Piece of Cloth), which introduced a new concept: the cutting of a garment in the shop at the customer's request.

Alfredo Häberli (b.1964 in Argentina) lives and works in Switzerland. His designs are disciplined and unfussy: 'many functions, much thought, few forms'. One of his particular areas of interest is air, synonymous with freedom and breathing. He draws inspiration from everyday life. A project of his can correspond to an idea, a sentence or an image. He was chosen as designer of the year at the exhibition 'Now 2004', where he showed his *Take a Line for a Walk* chair (2004, made by Moroso).
In England, Jasper Morrison presented his latest – and very simple –

The *ATM* (Advanced Table Module) desk, seen from underneath, reveals the lightness of its structure. Jasper Morrison, 2003. Manufactured by Vitra.

The *ATM* (Advanced Table Module) desk range, Jasper Morrison, 2003. Manufactured by Vitra.

range of desks, *ATM* (Advanced Table Module, 2003), at the new Vitra showroom. An aluminium screen that can be adjusted to different heights makes it possible to cut oneself off from one's surroundings. The critic Charles Boyer has described Morrison's design ethos as 'to produce everyday objects for everyone's use, make things lighter not heavier, softer not harder, inclusive not exclusive, generate energy, light and space'.

In Germany, the designer Konstantin Grcic joined the company Authentics. Run by Hans Maier-Aichen since 1980, Authentics produces kitchen and household articles that adhere to a minimal, economic and ecological design concept – for example, the *Leg Over* stool by Sebastian Bergne (1997) and the *Rondo* shopping bag by Hans Maier-Aichen. The company follows detailed aesthetic criteria and has a pricing policy. Konstantin Grcic created a presentation for Authentics at the Milan Furniture Exhibition of 1996. Although regarded as minimalist, his designs are not devoid of humour. He designed the *Pallas* table (2003), and *Diana* (2004) – a series of occasional tables made from folded steel and manufactured by Classicon. He also designed the *Chair One* chair (2003), made from aluminium and concrete and manufactured by Magis. His *Mayday* lamp was awarded the Compasso d'Oro in 2001, and is in the collection of the Musée d'Art Moderne in Paris.

Views of the interior design scheme of the Hi Hotel, Matali Crasset, Nice, 2003.

The work of the Belgian designer Maarten van Severen is austere and restrained. He has produced designs such as the *.03* polyurethane stacking chair, manufactured by Vitra, or the *LCP* transparent chaise longue, manufactured by Kartell and shown at the Cologne Furniture Exhibition in 2002. He was chosen as 'Designer of the Year 2001' at the Paris Furniture Exhibition. He designed the chairs for the library at the Pompidou Centre in Paris. His shelving units (2004), made from aluminium and polycarbonate, were manufactured by MarteenvanSeveren Meubelem.

Humour and empathy

Matali Crasset is a symbolic figure for the new generation in France. She has turned her face and haircut into a logo, using it systematically in her publicity. She is an object of fascination for the media. Her thinking is based on the belief that people must be helped to like, understand and meet one another. At the Milan Triennale in 1991, she exhibited her *Trilogie Domestique*: three diffusers, one of heat, one of light and one of water. She then worked with Denis Santachiara in Milan (1992), with Philippe Starck, who was in charge of the Thomson Multimedia project, and with Tim Thom Design, for whom she designed the personal stereo *O+O*, television sets, the radio cassette player *Don-O*, and many other audiovisual products (1993–7). She also designed some items for the German company Authentics, such as the *Empathic Chair*, a prototype for street furniture (1996); *W at hôm*, a prototype for domestic office

Quand Jim Monte à Paris, a 'hospitality column' (spare bed), Matali Crasset, 1998. Made by Domeau & Pérès.

furniture, funded by a Carte Blanche bursary (1997); and *Quand Jim Monte à Paris* (1997), a column that can be transformed into a bed for an unexpected guest. Matali Crasset set up her own studio in 1998. Among many other projects she designed a radio alarm clock for Lexon. After being awarded the Grand Prix of the international press by contemporary furniture critics for *Quand Jim Monte à Paris*, made by Domeau and Pérès (1999), she designed a collection of prototype furniture, *Glassex*, with Olivier Peyricot and Lisa White (1999).

Matali Crasset's main concern is the living space: 'There are many ways of rethinking the home by working on dead spaces or by building flexible micro-architectural elements'.

The home would be split up into layers at ground level: the relaxation area, the area for work, and so on. She exhibited *Table et Chaises Travesties* in 2000 at the Galerie Gilles Peyroulet. The covers can be hung up on the wall when they are not being used. In 1999 and 2000, she designed many objects: a self-generating lamp (giving out at night the light accumulated during the day); *Next to Me*, a chair for an office waiting area; *Il Capriccio di Ugo*, an armchair with armrests in the form of shelves; *Téo de 2 à 3*, a snoozing stool; and *Permis de Construire*, a sofa in the form of building blocks. Crasset designed the interior design scheme for the Hi Hotel, which opened in 2004 in Nice, in a 1930s building. For this she devised a space enlivened by flat washes of primary

Embryo Chair, armchair with tubular frame, polyurethane and neoprene, Marc Newson, 1988. Manufactured by Idée.

colours, structured by nine distinct concepts used across the 38 bedrooms, with original furnishing elements serving two purposes: for example, the bedhead is also a desk. The standard bedroom is a single space with no partitions: the bathroom is in the bedroom.

Another design group burst onto the Parisian scene: the Radi Designers (Recherche Autoproduction Design Industriel). Formed in 1992, this group has five members: Claudio Colucci (who set up an offshoot in Tokyo), Florence Doléac Sadler, Vincent Massaloux, Olivier Sidet and Robert Stadler. They play games with the seemingly obvious in everyday life, with the typologies of objects, products, furniture and gadgets. They manipulate codes, usages, techniques and forms, and enjoy inventing things.

Tavolino, a pedestal table with a carafe that plugs into its centre, and *Ray*, a stool made in the profile of Ray Eames, are produced by the Galerie Sentou. The Radi Designers aim to stimulate the senses and add a dash of humour to the world of furniture and objects, as demonstrated by their *Whippet Bench*, a life-size reproduction of a greyhound; their *Sleeping Cat* rug (1998), in which artificial flames, an electrically heated mat and a sleeping cat provide a real feeling of cosiness; or their *Coffee Drop Splash* (1998), a biscuit that looks like a splash. They were chosen as designers of the year at the Paris Furniture Exhibition in 2000.

The Australian designer Marc Newson (b.1963) can turn his hand to anything: furniture, seating, tables and lights; utensils, bottle openers, draining racks, flashlights, or a jug for a brand of pastis. He has adopted the 'Pod' (as in bean pod) as his favourite shape, using it in his very first design, the *Lockheed Chair* (1987). In 1995, he created *Bucky*, a dome composed of 'Pod' elements, in homage to the American designer Richard Buckminster Fuller. He has become a design star: his style is sexy, humorous and futuristic. However, machines that defy time and space are his true passion. He has designed the interiors of private jets such as the Falcon *900B*, bicycles for the Danish company Biomega, a concept car (the Ford *021C*), and a seat for Qantas (the *Skybed*). At the Fondation Cartier in 2004, he exhibited his 'ideal object', the *Kelvin 40* – a two-seater plane with an eight-metre wingspan and carbon wings grafted on to an aluminium fuselage.

A crisis in design?

In the preface to the catalogue of the Marc Newson exhibition at the Fondation Cartier (2004), the architect and aviation enthusiast Paul Virilio wrote: 'Newson reveals the crisis in design, which is now asking philosophical questions which art or architecture can no longer ask.' A crisis in design, or rather a crisis in marketing, for large companies rely on designers to add value to their products. 'The thing that has always driven me as a designer,' Marc Newson says, 'is feeling pissed off by the shitty stuff around me and wanting to make it better.' The Groninger Museum in the Netherlands organized a major retrospective of Newson's work in 2004. The English manufacturer Inflate re-introduced inflatable furniture, stressing the pleasure it gives, its rounded forms, its softness and lightness, and even coming up with a series of likable gnomes. The designers Marc and Michael Sodeau and Nick Crosbie produced egg-cups, lamps, bath plugs and salad bowls.

In the Netherlands, the group Droog Design was formed by rebel designers including Gijs Bakker and Renny Ramakers. Ramakers is the artistic director and co-founder of the group. She sees creative design as an exciting, exhilarating intellectual game, a reaction against form. 'Yes, beauty still means something ... It should always have a disturbing side.' In 1993, they exhibited at the Milan Furniture Exhibition, showing a chandelier by Rody Graumans consisting of 85 naked bulbs in the form of a bouquet. Richard Hutten, who works in wood, created a bench based on a reverse swastika. Tejo Remy attracted attention with the *Milk Bottle* light (1991).

Transversality

Éric Jourdan, the organizer of the Saint-Étienne Biennale in 2000 and 2002, studied at the Saint-Étienne School of Fine Arts. He designed the interior of the mayor's office there together with François Bauchet (1996). His career has been punctuated by rewarding collaborations:

Marie-Claude Beaud at the Fondation Cartier, for which he designed signs in 1989; Philippe Starck and Thomson, with whom he created the *Oyé Oyé* radio (1993); and Michel Roset, who manufactured his *Tolozan* chaise longue (2002). Jourdan's research relates to the continuity of spaces. In 1995, he started to focus on architectural furniture, going on to design his *Traversant* partition units (2000), exhibited at the Galerie Peyroulet. He presented *Espace Visionnage* at the VIA in 2002. Jourdan also designed the reception area furniture for the Maison Internationale at the Cité Universitaire (student halls of residence) in Paris (2004). The *Hyannis-Port* desk (2004, made by Cinna) is one of his small pieces of furniture that come in various sizes.

In 1994, Frédéric Ruyant set up Design Affairs. He produced furni-

Traversant partition and furnishing element made from painted MDF and varnished poplar, Éric Jourdan, 2000. Galerie Peyroulet.

ture for the new Issey Miyake showroom (1997). Funded by the VIA in 1998, he designed the *Doux Dream* polypropylene lamp, which combines enjoyment with intimacy and serves both as a lamp and a tidy.

He produced a collection of lights for Cinna (1998), and designed furniture for Ligne Roset (1999). His very minimalist approach reflects a desire for freedom. He makes it a principle not to become attached to objects or to feel nostalgic about them; he likes to change them and recycle them. His *Mobilier en Ligne, Ligne de Mobilier* continues a project started in 1999. He has designed *Dining Suite* and *Wood Corridor*, two new projects which result in a totally transformed space. He was awarded a Carte Blanche bursary by the VIA for his *Mobile Home* furniture (2003), and was commissioned by the VIA to design a medium-sized flat for the 2004 Furniture Exhibition. The space was decompartmentalized and rearranged.

All the rooms were interlinked, and furniture was not fixed in any one room: for example, the desk could slide in a single movement from the bedroom into the living-room. The project presented by Jean-Michel Policar at the Paris Furniture Exhibition in 2004 reveals the same desire for transversality: an adaptable furniture-unit-cum-area, designed for comfort, neither completely inside nor completely outside. Echoing this, the *Noon* light is powered by naturally provided energy (from photovoltaic sensors).

Interactive design

New technologies were introduced into design via architecture, which was already using three-dimensional tools. CD-ROMs with encyclopaedic potential were the initial teaching tools, but infatuation with them was short-lived. The expansion of the Internet, on the other hand, speeded up communication on a global scale. Dominique Mathieu is a new generation designer trained on Autocad and Silicon Graphics. Conceiving his projects virtually, he is primarily concerned with passing them on and making them into images. He designed *Clou* (2000), produced by the Galerie Gilles Peyroulet: the shadow cast on the ground by the pedestal table is materialized and serves as a base. He also designed an amazingly simple shelving unit (2004), made of folded steel and manufactured by FR66.

Digitization, information technology and new technologies

There have been huge strides in the development of technology. In 1990, the first digital photography system was introduced by Kodak. Then came the invention of the photo CD, the launch of Sony's MiniDisc (1992), the launch of the Viewcam camcorder by the Sharp Corporation (1992), and the launch in Japan of the DVD (Digital Versatile Disc, 1996), considered likely to become the universal platform for storing digital information. In 1997, plasma screens and giant flat screens were perfected. Kodak produced the first rewritable CD. Sony invented the flat tube for TV screens. The personal digital assistant *Avigo 10*, created by Texas Instruments, combined the functions of an electronic diary with a PC relay, and a great deal of information could be stored on it.

Among the new design objects of the 1990s, those associated with electronics and information technology occupied pride of place. 'Virtual reality' was accepted as a mode of representation, becoming important in the conception and modelling of design research. The computer came to play a crucial role in everyday life. The company Apple successfully adapted to developments in lifestyle, employing the skills of the designer Jonathan Ive, formerly a member of the Tangerine Studio. It was his brilliant idea to design a transparent computer in South Seas colours.

The first prototype came out in 1997, and the *iMac* went on sale in 1998. It was an instant success. Apple revolutionized the world of the microcomputer by launching a product that was not only a working tool but also, in a way, a fellow conspirator. 'New edge' in style, with integral handles, the transparent blue-tinted case revealed the internal components of the computer. Apple preserved the idea of user-friendliness while at the same time adding the factors of power, integration into a single unit, and practicality (ease of connecting with peripherals, and USB ports). Many products were brought out in the same range, consolidating the leading position of the Apple brand.

Prototypes and runs

Prototyping is a means of making models and prototypes quickly from a digital image of the object. The image is produced using CAD (computer-aided design), and is supported by a digital file containing all the information about the object in 3D. This technology has been developed since the 1980s in the fields of car, medical, aeronautical and object design. It is divided into two types of process: Stratoconception, which works by superimposing layers of photosensitive resin which is polymerized by means of a laser, and the LOM (Lamination Object Manufacturing) process, in which layers of powder (metal, nylon, ceramics, plastic, and so on) are superimposed. These different prototyping processes provide new possibilities for making one-off items or limited runs intended to go onto the market straight away.

Some designers have opted to use this technique to produce finished goods rather than prototypes. There has been a spate of projects recently. In 2003, the Italian designer Gabrielle Pezzini and three colleagues collaborated on *Made in China*, a collection of everyday consumer goods produced by means of stereolithography and made by the American company DSM Somos. The Finn Janne Kyttanen and the Dutchman Jiri Evenhuis exhibited the *Honey Bunns* stool, made from honeycomb resin, at the 2004 Furniture Exhibition in Cologne, while at the 2004 Milan

Furniture Exhibition the Belgian company Materialise showed the *.mgx* collection: lights devised by Kyttanen and Evenhuis. The Milanese firm Oneoff, founded in 2002, invited young designers to produce a collection called 'In Dust We Trust'. It took just a few hours' work to make these prototype objects (four hours for Apostolos Porsanidis's *Money* ring, 35 hours for the *Nopully* light fitting by Cristiana Giopato and Christopher Coombes). The pieces are coated with resin, but their delicate appearance makes them look as fragile as ceramics. This sophisticated technique of prototyping makes it possible to produce complex forms quickly and opens up new possibilities for contemporary design.

Digital Internet refrigerator *GR-D267DTU*, 2003. It is equipped with a television, a radio, a digital camera, messaging and Internet surfing – the ultra-sophistication of the household electrical appliances of tomorrow. Made by LG Électroménager.

Design-fiction

Conceptual design assigns a crucial role to technology. This powerful tool allows product typologies to be rethought. Styles can be mixed, and marginal trends become all-important. Conceptual design thus marks the end of the idea of an avant-garde. Thanks to montage and collage, past and future can coexist; contradiction brings richness to a design; chaos theory is taken up and reinterpreted. In terms of communications, human beings exist in a space that has been greatly altered by new technologies. They are developing virtual mobility, with the computer occupying a central place in their homes. They follow trends and fashions, and the home is being transformed into a zone of free inter-relational exchange.

Science fiction or reality

Reality is catching up with science fiction. The *Screenfridge* refrigerator (1999) by Electrolux was equipped with a computer and a touch screen. Predictions about future technologies and their repercussions on our daily lives can seem far-fetched, and yet some extraordinary design objects have already been produced: miniaturized audiovisual equipment, communication tools and computers in tiny containers, interactive washing machines, speaking kitchens, touch-controlled images projected onto all kinds of supports, virtual robots, and so on. We can use remote-control devices to lock and unlock car or house doors, turn on the light or the oven, fill the bath, and change television channels.

In 1983, Ugo La Pietra presented the telematic home of the future at the International Fair in Milan; it is already outmoded. Twenty years later, in 2003, the LG Électroménager system was unveiled at the Furniture Exhibition in Paris. In the midst of the digital domestic revolution, LG came up with the idea of 'neoculture', in which an intelligent digital living environment is used to make daily life easier. The leading company where digital technology is concerned, LG produced an ultra-sophisticated home automation network. The *Internet* refrigerator is equipped with TV, radio and digital camera functions, MP3, messaging and Internet surfing. It can be used as a network server, using the PLC (Power Line Communication) communication exchange protocol. It has a 15-inch monitor with an adjustable touch screen, four loudspeakers with digital sound, a digital camera, a microphone and a remote-control receiving assembly. It acts as a network server for other household electrical appliances. The *Internet* microwave oven gives access via the Internet to recipes and shopping advice which will help you manage your budget. It is also possible to connect to the LG website and programme the washing cycle of the washing machine or set the air-conditioning unit at an ideal temperature. This 'ideal life' linked to the Web represents a very state-of-the-art application of the new technologies, opening up dizzying possibilities. But will we become dependent on the Internet?

Conclusion

There is not one type of design but many, and these depend on the attitude towards the object and the project under consideration.

'Ugliness doesn't sell'

The history of design, which started in 1851 with the Great Exhibition in London, the first World Fair, is intimately bound up with developments in society. The Industrial Revolution provided the trigger, with the industrialization of the manufacturing process and mass production. Craftsmanship and hand-operated machinery gradually disappeared. The question of what products looked like and what the commercial impact of their appearance might be was explored in the USA. One answer was given by Raymond Loewy: 'Ugliness doesn't sell.' At the height of the economic crisis following the Wall Street crash, the products that sold best were those that were best designed. It was in the USA in the 1930s that a company and a designer first entered into a contract, thereby conferring a professional status on the designer.
Design in Europe went through a succession of ideologies: Arts and Crafts, the Werkbund, the Bauhaus, the UAM (Union des Artistes Modernes). Then, spurred on by the USA, industry became involved. In Italy in particular, industry benefited from a proper design policy. The company Olivetti and the furniture industry were very active. New plastic materials were used. German 'gute Form', the company Braun and the industrial aesthetics of such people as the French de-signer Roger Tallon all testified to the prevailing dynamism. Design swings back and forth like a pendulum, and at the end of the 1960s, there was a renewed questioning of design, with anti-design groups (Alchimia, Memphis) shaking up the design world and adopting a subversive attitude. Then, following the oil crisis, designers began to ask ethical questions related to the excesses of the consumer society. And finally, today, objects can be designed with the help of three-dimensional virtual images, which give free rein to the most utopian of projects.

Utilitarian objects

In essence, design focuses on the relationship between form and function, linking concept and design, project and form. The object is considered as a utensil, equipped with a programme controlling its use. It is appreciated and sought after for its practical value, its functionality, its convenience, its performance, its effectiveness or its ease of use. It is sought after for its quality–price, quantity–price or cheapness–safety ratio. Furthermore, the globalization of markets and the development of technology are leading to increasing market and product uniformity.

Creative designers are influenced by tradition, but they have often been trained in other countries. Mobility and versatility are among the major factors that have transformed design, disrupting established habits. With 300 shops in Japan, Europe and the USA, the Japanese company Muji offers a wide range of products: stationery, clothes, furnishing fabrics, utensils, crockery and furniture – good-quality, simple, unbranded products that are intended to blend in with every domestic interior. Founded in 1980, Muji has achieved great success with its natural, cellophane-wrapped products. Its strategy is to offer unbranded products carrying no logo.

Objects of desire

An object is a tool that enables each of us to achieve a purpose as part of an individual programme for living. In this respect it assumes the identity of the person using it. But above all, people expect the aesthetics of an object to enhance their lives emotionally: one way this can happen is through the pure spontaneous pleasure that an object provides.

Designers are trying more and more to understand and conceptualize new lifestyles. Matali Crasset creates a world of poetry. She thinks in terms of hospitality and intimacy, and possesses the quality of empathy: the intuitive ability to put herself in another person's place. Technology is becoming a means of stimulating people's imagination and desires. In one of the categories in the competition held at the exhibition 'Design for Europe 02' at Courtrai, some designers invented fairy-tale decors, taking us on a nostalgic trip back to childhood. They used items such as a survival kit, a folding shelter or a rocking chair. Projects like these are an ideal vehicle for the imagination.

High-tech objects

Ever more 'intelligent' materials have forced designers to re-examine the simple relationship between form and function. There is a huge, complex variety of materials; some are a cross between plants and high tech. Jean-Marie Massaud's *Horizontal Chair* is upholstered with a protein-based self-healing skin. Use is made of recyclable materials, high-performance composites, elastomers with a wide variety of textures or shock-resistant properties, textiles that conduct light or regulate temperature – all these are products that require a high level of technical expertise. Our conception of time has been altered: microprocessors contract time into tinier and tinier divisions. The organization of society and our understanding of the world have been turned upside down. The rapid development of information technology and telecommunications, as well as the use of the Internet, have opened up hitherto undreamt-of possibilities in our daily lives. The key word is 'convergence'. It is not just technologies that are being brought closer together but ways of using them too. The demand for mobile data services is expanding: 95% of those who own a mobile phone say they would like to have Internet access on it.

The company Panasonic foresees a world where it will be possible to watch television, listen to music and control household electrical appliances via a mobile phone. The Japanese manufacturer NEC is developing a set of tools with an interface allowing the user unlimited access to information, while at the same time protecting the natural environment. The tools include the *Ubiquitous Media-Chip Gumi* – microchips enclosed in transparent edible capsules. These can be connected to a drive and then eaten after the images and music they contain have been used. The German company Siemens has brought out a tri-band mobile phone in the shape of a pen, designed to be easier to use for those familiar with SMS. It is capable of translating handwritten words into SMS. The Dutch company Securfone is planning to launch a telephone with a simplified interface for older people.

Home automation is sufficiently developed for the *e-Home*, the online intelligent home which will have its various electronic systems linked to the net. Siemens, Bosch, Grundig, LG and Motorola have adopted a common standard, the EIB, thus making possible the remote control of most of the functions of the home and thereby saving energy. The German research institute Fraunhofer foresees the widespread use of domestic robots that are capable of simplifying household chores. *Aibo*, Sony's little Japanese robot dog, has been on the market since 2001. Ministries of Education in various countries support this interdisciplinary research work on the interaction between man and technology. The new technologies will allow people – both in their private and in their professional spheres – to control and use machines by simple means such as spoken commands or gestures.

The eco-object

Designers have to be responsible for their creations. They are under an obligation to think ahead, and foresee what the future life of their products will be. Design is part of the service sector, not just a branch of the arts. All products have an impact on the environment; all products consume and pollute. Designers must take account of this reality right from the conception and development phase. The way the European market is evolving is redefining industrial design. Designers are gradually taking on board the issues of energy consumption, typology of materials and new uses at the different stages of a product's life, and are taking account of packaging, consumables and point-of-sale advertising. Designers have to minimize and optimize.

Our daily lives are swamped by an overabundance of products. Our economy is still based on obsolescence and the idea that things will wear out. The transition from a consumer society to a user society would have to take place against a background of sustainable development, and environmental, social and ethical considerations would have to be integrated into economic planning.

We are currently witnessing the development of environmental management. The Grand Prix in the 'Objects for Survival and Outdoors' category of the exhibition 'Design for Europe' at Courtrai in 2002 was awarded to an object designed with materials and a technique associated with 'found objects', which are normally scorned by the design world. Created by the Belgian designer Sylvain Willenz and the English designer James Carrigan, *Dr Bamboozle* is a seat made from bamboo canes, roughly put together and impregnated with rubber to form a seat. Human beings are the guardians of the biosphere. They have to adopt an ethical attitude and be aware of all the consequences of their acts.

The 21st-century designer

As the expression of an industrial society and its materialistic influence, 20th-century design was centred on two main factors: material and energy. The age we are entering is more centred on time and living. Human beings are now able to simulate life and create new beings at the interface between biology and electronics. We are producing virtual creatures, descendants of the Tamagotchis so casually tended to by Japanese and European children. We are coming up with new hybrids created by genetic and electronic manipulation. This new knowledge and these new technological possibilities will feed 21st-century thinking. A software designer can develop his design according to how those using it react, thus combining the skills of engineer, scientist and creative designer. Up until now, designers have been engineers, interior designers, furniture designers, and so forth. They have known how to collect, classify, question, develop and create. But the 21st-century designer will not be able to operate without an in-depth scientific education which draws on the cognitive sciences, biology and information technology.

Appendices

Index

A

B

C

D

E

F

G

H

I

J

K

L

M

N

O

P

Q

R

S

T

U

V

W

Y

Z

Bibliography

Journals of modern design

Blueprint
Design Week
ID
The Journal of Design History
The Journal of Visual Culture

General sources

Dormer, P, *Design Since 1945*, Thames and Hudson, London, 1993
Dormer, P, *The Meanings of Modern Design: Towards the Twenty-First Century*, Thames and Hudson, London, 1990
Fiell, C and Fiell, P, *Modern Furniture Classics, Postwar to Post-modernism*, Thames and Hudson, London, 2001
Forty, A, *Objects of Desire: Design and Society 1750–1980*, Thames and Hudson, London, 1986
Heskett, J, *Industrial Design*, Thames and Hudson, London, 1985
Julier, G, *The Thames and Hudson Dictionary of Design Since 1900*, Thames and Hudson, London, 2004
Lubbock, J, *The Tyranny of Taste: The Politics of Architecture and Design in Britain 1550–1960*, Yale University Press, New Haven and London, 1995
McCarthy, F, *British Design Since 1880: A Visual History*, Lund Humphries, London, 1982
McDermott, C, *Design Museum Book of Twentieth Century Design*, Carlton Books, London, 1999
Sparke, P, *A Century of Design: Design Pioneers of the 20th Century*, Mitchell Beazley, London, 1999
Sparke, P, *Design in Context*, Bloomsbury, London, 1987
Walker, J A, *Design History and the History of Design*, Pluto Press, London, 1989
Whitechapel Art Gallery, London, *Modern Chairs 1918–1970*, exhibition catalogue, Lund Humphries, London, 1971
Woodham, J, *Twentieth-Century Design*, Oxford University Press, Oxford, 1997

http://www.designmuseum.org

Chapter 1

Anderson, S, *Peter Behrens and a New Architecture for the Twentieth Century*, MIT Press, Cambridge, MA, 2002
Crawford, A, *Charles Rennie Mackintosh*, Thames and Hudson, London, 1995

Davidson, E, *Ruskin and his Circle*, exhibition catalogue, Arts Council Gallery, London, 1964
Duncan, A, *Art Nouveau*, Thames and Hudson, London, 1994
Greenhalgh, P (ed), *Art Nouveau 1890–1914*, Victoria and Albert Museum, London, 2000
Heskett, J, *Design in Germany 1870–1918*, Trefoil, London, 1986
Morris, W, *News From Nowhere and Selected Writings and Designs*, edited by Asa Briggs, Penguin, Harmondsworth, 1984
Pevsner, N, *Pioneers of Modern Design from William Morris to Walter Gropius*, Penguin, Harmondsworth, 1991
Stedelijk Museum, Amsterdam, *Industry and Design in the Netherlands 1850–1950*, Coronet Books, Philadelphia, 1986

Chapter 2

Baird, G, *Alvar Aalto,* Thames and Hudson, London, 1970
Banham, R, *Theory and Design in the First Machine Age*, Architectural Press, London, 1980
Bayer, H et al, *Bauhaus 1919–1928*, Museum of Modern Art, New York, 1938
Fiedler, J and Feieraband, P (eds), *Bauhaus*, Könemann, Cologne, 1999
Frampton, K, *Le Corbusier*, Thames and Hudson, London, 2001
Giedion, S, *Mechanization Takes Command*, Oxford University Press, Oxford, 1948
Giedion, S, *Space, Time and Architecture*, 5th edn, Harvard University Press, Cambridge, MA, 1967
Hayward Gallery, London, *Le Corbusier, Architect of the Century*, exhibition catalogue, Arts Council of Great Britain, 1987
Le Corbusier, *Towards a New Architecture*, translated by Frederick Etchells, Architectural Press, Oxford, 1991
Naylor, G, *The Bauhaus Reassessed: Sources and Design Theory*, Herbert Press, London, 1985
Overy, P, *De Stijl*, Thames and Hudson, London, 1991
Peto, J and Loveday, D (eds), *Modern Britain 1929–1939*, exhibition catalogue, Design Museum, London, 1999
Tretiack, P, *Raymond Lowey*, Rizzoli, New York, 2000
Whitford, F, *Bauhaus*, Thames and Hudson, London, 1984

Chapter 3

Kicherer, S, *Olivetti*, Rizolli, New York, 1990
Lewis, E, *Great IKEA!: A Brand for All the People*, Cyan Books, London, 2005
Neuhart, J, *The Chair*, Design Museum, London,1989
Neuhart, J, *Eames Design: the Work of the Office of Charles and Ray Eames*, Ernst & Sohn, Berlin, 1989

Sparke, P, *Italian Design: 1870 to the Present*, Thames and Hudson, London, 1988

Sulzer, P, *Jean Prouvé Highlights, 1917–1944*, Birkhauser Verlag AG, Zürich, 2002

Wilson, R G, *The Machine Age in America*, exhibition catalogue, The Brooklyn Museum in association with Harry N Abrams, New York, 1986

Maguire, P J and Woodham, J M (eds), *Design and Cultural Politics in Postwar Britain: The 'Britain Can Make It' Exhibition of 1946*, exhibition catalogue, Leicester University Press, London, 1997

Chapter 4

Ambasz, E (ed), *Italy: The New Domestic Landscape*, exhibition catalogue, Museum of Modern Art, New York, in collaboration with Centro Di, Florence, 1972

Centre Georges Pompidou, Paris, *Design Français 1960–1990: Trois Déciennes*, exhibition catalogue, Agence pour la promotion de la création industrielle, 1988

Clifton-Mogg, C, *Conran's Habitat*, Crown Publications, London, 1991

Cook, P (ed), *Archigram*, Princeton Architectural Press, Princeton, 1999

Frampton, K, *Modern Architecture: A Critical History*, 3rd edn, Thames and Hudson, London, 1992

Gateshead, Baltic Arts Centre, *Archigram*, 2004

Jencks, C, *Modern Movements in Architecture*, 2nd edn, Penguin, Harmondsworth, 1985

Phillips, B, *Conran and the Habitat Story*, Weidenfeld and Nicolson, London, 1984

Whiteley, N, *Pop Design: Modernism to Mod*, Design Council, London, 1987

Chapter 5

Foster, H (ed), *Postmodern Culture*, Pluto Press, London, 1985

Jencks, C, *What is Postmodernism?*, 4th edn, Academy Editions, London, 1996

McDermott, C, *Street Style: British Design in the 1980s*, Design Council, London, 1987

Sottsass, E, *Sottsass Associates*, Rizzoli, New York, 1988

Venturi, R, *Complexity and Contradiction in Architecture*, 2nd edn, Museum of Modern Art, New York, 1977

Chapter 6

Centre Georges Pompidou, Paris, *Design Japonais 1950–1995*, exhibition catalogue, Centre Georges Pompidou, Paris, 1996

Coates, N, *Nigel Coates: The City in Motion*, Rizzoli, New York, 1989

Collings, M, *Ron Arad*, Phaidon Press, London, 2003

Fitoussi, B, *Memphis*, Thames and Hudson, London, 1998

Pernodet, P, *Luigi Colani*, Dis Voir, Paris, 2001

Rees, H, *Metropolis: Tokyo Design Visions*, Design Museum, London, 1991

Sweet, F, *Philippe Starck: Subverchic Design*, Thames and Hudson, London, 1999

Chapter 7

Hirst, D (ed), *Damien Hirst, The Agony and the Ecstasy: Selected Works from 1989 to 2004*, Electa Napoli, Naples, 2005

Lloyd-Morgan, C, *Marc Newson*, Thames and Hudson, London, 2003

Morrison, J, *Jasper Morrison: Everything But the Walls*, Lars Müller, Baden, 2002

Williams, Gareth, *Crasset Matali – Design & Designer 006,* Pyramyd, Paris, 2004

http://www.apple.com

Photo Credits

p 8 : HUNTERIAN MUSEUM • p 11 : AKG • p 13 : VITRA DESIGN MUSEUM • p 15 : Cheltenham Art Gallery and Museums/BRIDGEMAN - GIRAUDON • p 16 : Ph Christine Smith © Archives Larbor • p 17 : BRIDGEMAN - GIRAUDON • p 18 : The Fine Art Society, London/BRIDGEMAN - GIRAUDON • p 19 : William Mooris Gallery/BRIDGEMAN - GIRAUDON • p 21 : Archives Larbor • p 22 : RMN • p 24 : L Maisant/Corbis • p 27 : R G Ojeda/RMN© ADAGP Paris 2004 • p 29 : H Lewandowski/RMN • p 30 : J Schormans RMN • p 31 : Klgnatiadis/RMN • p 35 : T Roche/MUSÉE D' ART MODERNE SAINT-ÉTIENNE • p 36 : E Lessing/AKG • p 38 : AKG • p 41 : Bethnal Green Museum/BRIDGEMAN - GIRAUDON • p 42 : Deutsches Technikmuseum Berlin • p 44 : E Lessing/AKG • p 47 : Archives Larbor© ADAGP, Paris 2004 • p 48 : Archives Larbor©ADGP, Paris 2004 • p 50 : Photothèque des collections du Centre Georges-Pompidou/Musée National d'art moderne, Paris© Marcel Duchamp estate/ADAGP, Paris 2004 • p 51 : Ph G Lepkowski © Archives Nathan © ADAGP, Paris 2004 • p 53 : BAUHAUS ARCHIV • p 54 : Gunter Lepkowski/BAUHAUS ARCHIV BERLIN • p 55 : T Dix/VITRA DESIGN MUSEUM/ ADAGP, Paris 2004 • p 56 : Ph © Bauhaus Archiv - Archives Larbor • p 57 : AKG • p 59 : Fondation Le Corbusier/ADAGP, Paris 2004 • p 61 : Fondation Le Corbusier/ADAGP, Paris 2004 • p 62 : CENTRE GEORGES-POMPIDOU/Bibliothèque Kandinsky • p 63 : TDIX/VITRA DESIGN MUSEUM • p 64 : Ullstein Bild/AKG © ADAGP, Paris 2005 • p 66 : Yves Bresson/MUSEE D ART MODERNE SAINT-ÉTIENNE • p 66 : CENTRE GEORGES-POMPIDOU/Bibliothèque Kandinsky • p 69 : Museum of Fine Arts, Houston/BRIDGEMAN - GIRAUDON • p 70 : CORBIS • p 72 : Bettmann/CORBIS • p 73 : Condé Nast Archive/CORBIS • p 77 : ARTEK • p 78 : ARTEK • p 80 m : CENTRE GEORGES-POMPIDOU/Bibliothèque Kandinsky • p 80 top : MOMA/SCALA • p 80 bottom : T Dix/VITRA DESIGN MUSEUM • p 82 : AKG • p 84 : Fritz Hansen • p 87 : T Dix/VITRA DESIGN MUSEUM/ADAGP, Paris, 2004 • p 88 : Y Bresson/MUSEE D ART MODERNE SAINT-ÉTIENNE • p 89 : T Dix/VITRA DESIGN MUSEUM • p 90 : Herman Miller • p 93 : Herman Miller • p 94 : CENTRE GEORGES-POMPIDOU/Bibliothèque Kandinsky • p 96 : SONY FRANCE • p 98 : rights reserved • p 101 : VITRA DESIGN MUSEUM • p 102 : Y Bresson/MUSEE D ART MODERNE SAINT-ÉTIENNE • p 103 : IKEA • p 105 : A Platen/AKG • p 107 : rights reserved • p 108 : CENTRE GEORGES- POMPIDOU/Bibliothèque Kandinsky • p 110 top : Coll Enrico Maltoni • p 111 : M Panicucci/ Art Direction,Max Pinucci • p 113 : CENTRE GEORGES -POMPIDOU/Bibliothèque Kandinsky • p 115 : © Marc Domage/Tutti-Courtesy Jousse Entreprise/ADAGP, Paris 2004 • p 116 : ROGER-VIOLLET • p 117 : ROGER-VIOLLET • p 118 : RENAULT COMMUNICATION • p 120 : CENTRE GEORGES -POMPIDOU/Bibliothèque Kandinsky • p 124 : rights reserved • p 126 : Herman Miller • p 128 : T Dix/VITRA DESIGN MUSEUM • p 130 bottom : Bettmann/CORBIS • p 131 : HABITAT • p 133 : rights reserved • p 134 : rights reserved • p 136 : Y Bresson/DR/MUSEE D ART MODERNE SAINT-ÉTIENNE • p 137 : rights reserved • p 139 : ZANOTTA spa • p 140 : VITRA DESIGN MUSEUM • p 141 : T Dix/VITRA DESIGN MUSEUM • p 142 : ZANOTTA spa • p 144 : Bang Olufsen • p 145 : CENTRE GEORGES-POMPIDOU/Bibliothèque Kandinsky • p 146 : Vico Magistretti for ARTEMIDE • p 148 : ZANOTTA spa • p 153 : H Hansen/VITRA COLLECTIONS • p 154 : VICTORIA AND ALBERT MUSEUM • p 155 : VITRA DESIGN MUSEUM • p 157 : P Magnon/Collection FRAC Centre, Orléans • p 158 : VITRA DESIGN MUSEUM • p 159 : T Dix/VITRA DESIGN MUSEUM • p 160 : TDix/VITRA DESIGN MUSEUM • p 161 : Alessandro MENDINI • p 162 : Y Bresson/MUSEE D' ART MODERNE SAINT-ÉTIENNE • p 163 : T Dix/VITRA DESIGN MUSEUM • p 164 : Fonds national d'art contemporain- Dépôt au MUSEE D' ART MODERNE -ÉTIENNE • p 166 : ALESSI France • p 168 : Starck Network • p 171 : Editions du Regard/rights reserved/ADAGP/Paris, 2004 • p 173 : CENTRE GEORGES- POMPIDOU/Bibliothèque Kandinsky/ADAGP, Paris 2004 • p 175 : T Dix/VITRA DESIGN MUSEUM • p 176 : C Recoura/LVDR/DIAPORAMA • p 178 : Richard Sapper pour ARTEMIDE • p 181 : CENTRE GEORGES- POMPIDOU/Bibliothèque Kandinsky • p 182 : DYSON • p 183 : M Lelièvre/Studio Bleu/Starck NetworK • p 184 : rights reserved • p 186 : MUSEE D' ART MODERNE SAINT-ÉTIENNE • p 187 : Museum of Fine Arts, Houston/BRIDGEMAN - GIRAUDON • p 190 : TOM DIXON • p 195 : Galerie Peyroulet/Paris • p 196 : François Azambourg/Editions Domeau et Pérès • p 199 : THOMSON MULTIMEDIA • p 201 : Rex Pictures/SIPA PRESS • p 203 : Martin SZEKELY • p 204 : M Lelièvre/Studio Bleu/ Stark Network • p 206 : Editions VIA • p 207 : Apple/AAR/SIPA PRESS • p 208 : Martin SZEKELY • p 209 : VITRA COLLECTIONS • p 210/211 : VITRA COLLECTIONS • p 213 : Lorenvu/SIPA PRESS • p 214 : P Gries/Domeau et Pérès • p 215 : Marc NEWSON Ltd/www.marc-newson.com • p 217 : Galerie Peylouret • p 220 : LG Electroménager •